QUANTIFIERS, DEDUCTION, AND CONTEXT

CSLI
Lecture Notes
No. 57

QUANTIFIERS, DEDUCTION, AND CONTEXT

edited by
Makoto Kanazawa, Christopher J. Piñón,
&
Henriëtte de Swart

CSLI Publications

CENTER FOR THE STUDY OF
LANGUAGE AND INFORMATION
STANFORD, CALIFORNIA

Copyright ©1996
CSLI Publications
Center for the Study of Language and Information
Leland Stanford Junior University
Printed in the United States
99 98 97 96   5 4 3 2 1

Library of Congress Cataloging-in-Publication Data

Quantifiers, deduction, and context / edited by Makoto Kanazawa, Christopher
J. Piñón & Henriëtte de Swart
        p.   cm. — (CSLI lecture notes ; no. 57)
   "The present volume is an outgrowth of, but by no means faithfully
represents, the second CSLI Workshop on Logic, Language, and Computation
… held at the Center for the Study of Language and Information … on June
4-6, 1993" –Pref.
   Contents: The context-dependency of implicit arguments / Cleo Condoravdi
and Jean Mark Gawron – A deductive account of quantification in LFG / Mary
Dalrymple … [et al.] – The sorites fallacy and the context-dependence of vague
predicates / Kees van Deemter – Presuppositions and information updating /
Jan van Eijck – Indefeasible semantics and defeasible pragmatics / Megumi
Kameyama – Resumptive quantifiers in exception sentences / Friederike
Moltmann – (In)definites and genericity / Henriëtte de Swart.
   Includes bibliographical references and index.
   ISBN 1-57586-005-8. — ISBN 1-57586-004-X (pbk.)
   1. Language and logic – Congresses. 2. Grammar, Comparative and
general–Quantifiers–Congresses. 3. Semantics–Congresses. 4. Context
(Linguistics)–Congresses. I. Kanazawa, Makoto, 1964– . II. Piñón, Christopher
J., 1963– . III. Swart, Henriëtte de, 1961– .IV. CSLI Workshop on Logic,
Language, and Computation (2nd : 1993) V. Series.
P39.Q36   1995   95-26587
401–dc20   CIP

CSLI was founded early in 1983 by researchers from Stanford University, SRI
International, and Xerox PARC to further research and development of integrated
theories of language, information, and computation. CSLI headquarters and CSLI
Publications are located on the campus of Stanford University.

CSLI Lecture Notes report new developments in the study of language, information, and
computation. In addition to lecture notes, the series includes monographs, working
papers, and conference proceedings. Our aim is to make new results, ideas, and
approaches available as quickly as possible.

♾The acid-free paper used in this book meets the minimum requirements of the
American National Standard for Information Sciences—Permanence of Paper for Printed
Library Materials, ANSI Z39.48-1984.

Contents

Preface

The papers collected here exemplify diverse approaches to the three central themes in the study of language and logic which make up the title of the book. Reflecting the interdisciplinary nature of the field, the concerns of the seven papers span natural language semantics, computational linguistics, and philosophical logic.

Since the early eighties, quantification in natural language has been at the center of attention for linguists and logicians interested in semantics—so much so that it has formed a subfield of its own, known as Generalized Quantifier Theory. The papers by de Swart and by Moltmann are examples in this subfield of linguistic analyses aided by logical machinery. De Swart develops a unified treatment of generic and non-generic readings of definite and indefinite NPs, analyzing quantificational adverbs as expressing quantification over events and using a dynamic interpretation of indefinite NPs. Moltmann shows that certain sentences with exception phrases essentially involve polyadic quantification, giving perhaps the clearest evidence found so far that quantification in natural language is not confined to monadic quantification.

Dalrymple, Lamping, Pereira, and Saraswat propose a treatment of quantifier scope ambiguity and the interactions between scope and bound anaphora in a semantically interpreted Lexical-Functional Grammar. They use linear logic as a 'glue language' to mediate functional structures of LFG and semantic interpretations. In their account, the interpretation process corresponds to a formal deduction in linear logic.

The subjects of the remaining papers all have to do with the notion of context in one way or another. Condoravdi and Gawron characterize the type of context-dependency involved in the interpretation of relational predicates with implicit arguments such as *win* and *local* and give a unified account of their readings. The contribution by van Deemter offers a solution to the Sorites Paradox, arguing that the context-dependency of the relevant vague predicate makes the crucial premise in the sorites argument ambiguous, and that confounding of the two readings is responsible for the fallacy. Van Eijck uses dynamic modal logic to analyze presupposition failures based on the idea that left-to-right processing of a piece of language, whether it is a computer program or a natural language text, involves a dynamic updating of contexts or information states. The paper by Kameyama tackles the issue of how discourse context controls the space of interacting preferences in discourse interpretation. Assuming a discourse processing architecture in which the grammar rules are indefeasible and the pragmatic principles are defeasible, she shows how independently motivated default preferences interact in the interpretation of intersentential pronominal anaphora.

Like its predecessor (*Dynamics, Polarity, and Quantification*, edited by Kanazawa and Piñón), the present volume is an outgrowth of, but by no means faithfully represents, the SECOND CSLI WORK-SHOP ON LOGIC, LANGUAGE, AND COMPUTATION, which was held at the Center for the Study of Language and Information (CSLI) on June 4–6, 1993. The workshop was organized by the editors of this volume together with Johan van Benthem and Stanley Peters. Financial support was provided by CSLI under the project titled 'The Relational Theories of Language as Action'. This support is gratefully acknowledged.

Makoto Kanazawa
Christopher Piñón
Henriëtte de Swart

Contributors

CLEO CONDORAVDI is visiting assistant professor at the University of Texas at Austin. Her main interests are in formal semantics and pragmatics and in the interface of syntax and semantics.

MARY DALRYMPLE is a researcher in the Natural Language Theory and Technology group at Xerox Palo Alto Research Center and a Consulting Assistant Professor in the Department of Linguistics at Stanford University. Her recent work focuses on the syntax-semantics interface.

KEES VAN DEEMTER works at the Institute for Perception Research (IPO) in Eindhoven, The Netherlands. IPO is jointly governed by Philips Research Laboratories and Kees' field of research is formal semantics of natural language. His current interests include the logic of ambiguity and vagueness, and more recently the use of semantics in the generation of prosodically acceptable speech.

JEAN MARK GAWRON is a research linguist at SRI International. His research interests include the syntax semantics interface, lexical semantics, and the semantics of anaphora, quantification, and comparatives.

MEGUMI KAMEYAMA is a research scientist at the Artificial Intelligence Center of SRI International in Menlo Park, California. She received a Ph.D. in linguistics and an M.S.C.S in computer science from Stanford University. Her main research areas are computational discourse models, logical theory of discourse pragmatics, grammar-pragmatics interface, and translation models.

JOHN LAMPING received his Ph.D. in Computer Science from Stanford University. His research focuses on developing expressive computational languages.

FRIEDERIKE MOLTMANN received her Ph.D. from MIT. Her research areas include formal semantics, the syntax-semantics interface and the philosophy of language.

FERNANDO PEREIRA heads the Information Extraction Research Department at AT&T Bell Laboratories, Murray Hill, New Jersey. His recent work has included finite-state models of speech recognition, statistical models of text, and deductive approaches to syntax and semantics.

VIJAY SARASWAT is a member of the Research Staff at Xerox Palo Alto Research Center.

HENRIËTTE DE SWART is assistant professor in linguistics at Stanford University. Her research interests include natural language semantics and pragmatics with a special focus on tense and aspect, quantification and dynamic theories of meaning.

1

The Context-Dependency of Implicit Arguments

Cleo Condoravdi and Jean Mark Gawron

1 Introduction

Context-dependency is traditionally taken to involve the context of utterance and meanings are specified as functions from contexts to content (Cresswell 1973, Kaplan 1989a, Lewis 1981, Stalnaker 1972).[1] Context-dependent linguistic expressions are assumed to comprise a small inventory of closed-class elements, such as tense, indexicals, deictic and demonstrative pronouns, and context determines content to the extent that it supplies values for these expressions. Empirical demands and theoretical developments, however, have greatly expanded the notion and scope of context-dependency. With the advent of dynamic semantics, for example, context has been construed in such a way as to incorporate information derived from previous linguistic discourse[2] and context-dependency has been expanded to include pronominal anaphora.

Mitchell (1986) and Partee (1989) have demonstrated that context-

---

[1] We are grateful to the members of the Indexicals Group at CSLI, including David Israel, Geoff Nunberg and John Perry, for the many stimulating seminar discussions that led to this paper. We had the opportunity to present this work first at the second workshop on Language, Logic and Computation in June 1993, and subsequently at the Joint Research Conference at Waseda University, and at colloquia at Stanford and Berkeley. We would like to thank the audiences at these events for their helpful comments. Thanks also to Tony Davis, David Israel, François Récanati for comments on a previous written version. Special thanks are due to Kees van Deemter and an anonymous reviewer for very thorough and detailed comments that substantially affected the final shape of this paper.

[2] The idea goes back to Stalnaker (1978) where the propositional content of previous linguistic discourse is available as a contextual parameter. DRT and dynamic semantics provide more fine-grained conceptions of context so as to account for intersentential and sentence-internal anaphora.

Quantifiers, Deduction, and Context
Makoto Kanazawa, Christopher Piñón, and Henriëtte de Swart, editors
Copyright © 1996, CSLI Publications

dependence extends even beyond the pronominal domain to encompass a large open class of contentful linguistic expressions.[3] Relational predicates such as *win, local, imminent* can be construed with an implicit argument and in order for them to be interpretable the context, linguistic or extra-linguistic, must supply a certain kind of information.[4] As observed by Mitchell and Partee, implicit arguments of such predicates have deictic, discourse anaphoric and bound variable readings. Given the similarity of implicit argument with overt pronominal elements in the type of context-dependency they exhibit and their range of readings, Partee (1989) argued for the need to extend the analyses of pronominal anaphora so as to cover the entire class of context-dependent expressions and for the need to integrate indexicality with variable binding.

In this paper we will assume that contentful context-dependent expressions are relational predicates with at least one implicit argument.[5] Our aim is to give a unified account of the readings of their implicit arguments and to characterize the type of context-dependency involved. We claim that context-dependent predicates are associated with familiarity presuppositions with respect to their implicit arguments and that these presuppositions account for the particular kind of context-dependency and for the readings of implicit arguments. In section 2 we distinguish implicit argument predicates that are crucially context-dependent from those that are not and describe their range of readings. In section 3 we provide an analysis of the anaphoric readings of implicit arguments. In section 4 we integrate the anaphoric and indexical readings of implicit arguments.

2 The Problem of Implicit Arguments

2.1 Types of Implicit Arguments

It is well-known that implicit arguments of verbal predicates can be interpreted in at least two ways (see, e.g., Fillmore 1969, 1986, Shopen 1973, Thomas 1979, Fodor & Fodor 1980, Dowty 1981). The implicit arguments of one type of predicate, such as *eat*, get an existential interpretation, whereas the implicit arguments of another type of predicate, such as *apply*, get an anaphoric interpretation.[6] This contrast is exemplified in (1).

---

[3]Mitchell (1986) calls them *perspectival expressions*.

[4]For a different perspective see Perry (1986).

[5]Partee considered this possibility but opted for a different one, whereby an expression of this sort indexes to a context rather than taking an extra argument. Indexing to contexts is not an easily available option within non-representational theories of context, such as the one we will be assuming.

[6]This distinction corresponds to Fillmore's (1986) distinction between indefinite and definite null complements, Shopen's (1973) distinction between indefinite and definite ellipsis and Thomas's (1979) distinction between nonrealization and ellipsis.

(1)   a. There was a piece of bread on the table but John didn't eat.
       b. There was a good job available here but Fred didn't apply.

(1a) implies that John didn't eat anything, whereas (1b) implies simply that Fred didn't apply for the good job that was available here (although he may have applied for other jobs).

This difference in interpretation turns out to be a consequence of a more basic constrast: predicates like *apply* depend on the context in an essential way, whereas predicates like *eat* do not. For this reason, the two types of predicates behave differently when used without any prior linguistic context. Predicates with existential implicit arguments are perfectly acceptable in such contexts, as in (2a), whereas predicates with anaphoric implicit arguments are infelicitous, as in (2b).

(2)   a. I painted last week.
       b. # I applied last week.

Another consequence of the inherent context-dependence of anaphoric implicit arguments is that they can take wide scope with respect to other quantifiers in the sentence, as can be seen in (3b) and (1b). Implicit arguments with an existential interpretation, on the other hand, take narrow scope with respect to all sentential operators, as can be seen in (3a) and (1a).[7]

(3)   a. I have been painting all week.
       b. I have been calling all week.

Predicates of both types may impose selectional restrictions on their implicit arguments, so this property crosscuts the basic context-dependent vs. non-context-dependent classification:

(4)   a. We need a lot of bricks for the construction job but no one is
          baking these days.
       b. The most attractive prize was a 20 pound turkey. John won.

Bake construed with an implicit argument, receiving an existential interpretation, can only be about the baking of bread or other baked goods, and not, for example, about roasts, or bricks. Similarly, the implicit argument of *win*, which gets an anaphoric interpretation, is understood to be a contest-type of entity rather than a prize-type of entity.

Although verbal predicates with implicit arguments have received the most attention, predicates of any syntactic category can be construed with an implicit argument. In (5) we give some examples of

---

[7]This contrast was originally pointed out by Fodor & Fodor (1980), who, moreover, claimed that anaphoric implicit arguments take exclusively wide scope. Narrow scope readings of anaphoric implicit arguments, however, are available, as shown already by Dowty (1981). We discuss them in detail in later sections.

predicates construed with a context-dependent implicit argument. For clarity, the implicit argument predicate is italicized and the necessary information for its interpretation is supplied in the same sentence.

(5)   a. He observed the ceremony from ten yards *away*/*across* the street.

      b. Lee sold his Vermeer after Marion bought a *similar* painting.

      c. When they put a locked gate on the road to the beach, Sal no longer had *access*.

The term 'non-context-dependent' should not be taken to imply that context does not in any way affect the way we understand non-context-dependent implicit arguments. In a particular context we may draw certain inferences on the basis of relevance considerations, discourse relations between sentences, etc., so as to derive additional information about the implicit argument. For instance, the discourses in (6a) and (6b) both imply that I have been baking pastries for the party.

(6)   a. We needed a lot of pastries for the party.
        I have been baking all week.

      b. I have been baking all week.
        We needed a lot of pastries for the party.

The crucial difference between the two types of predicates is that the context-dependency of context-dependent predicates is built into their *meaning*. As a result, (*i*) infelicity arises whenever the context cannot provide an element for their interpretation ((2a) vs. (2b)) and (*ii*) there are restrictions on what elements in a discourse can supply a value for an implicit argument, due to the way semantic interpretation works. As a result, it is the *prior* linguistic discourse that determines the felicity and interpretation of context-dependent implicit arguments. The order between two sentences makes a critical difference since the information contributed by the first is part of the context relative to which the second is interpreted but not vice versa. (7) and (6) contrast for this reason.

(7)   a. An explosives warehouse on the other side of town exploded yesterday. A nearby bar was seriously damaged.

      b. A nearby bar was seriously damaged. An explosives warehouse on the other side of town exploded yesterday.

(7a) and (7b) differ in that the location needed for the interpretation of *nearby* can be provided by the NP *an explosives warehouse* in (7a) but not in (7b). Hence (7a) and (7b) give rise to different inferences about the strength of the explosion in each case.

One may legitimately ask whether the lexical semantics of a given predicate determines in some way whether its implicit argument is given an existential or an anaphoric interpretation. Hopefully, the answer will turn out to be 'yes' but nobody has managed to show that[8] and we will not concentrate on this question here. Even if there exists such a connection, one still has to specify what the context-dependency amounts to and how it affects and is affected by the interpretation of the whole sentence. This is what we will concentrate on in this paper.

2.2 The Readings of Implicit Arguments

The interpretation of context-dependent predicates depends on circumstances of their use and, in order to be fixed, the context of utterance, the previous linguistic discourse, or the sentence in which they occur must supply information to fix the value of their implicit argument. For example, to determine whether there is a bar verifying the sentences in (8), we must have a specified location relative to which the bar is local. That location can be provided in a variety of ways depending on the context in which the predicate *local* appears.[9]

(8)   a. A local bar is selling cheap beer.
        b. A reporter for the *Times* got seriously drunk.
           A local bar was selling cheap beer.
        c. Every sports fan watched the Superbowl in a local bar.

In (8a) the relevant location is provided by the context of utterance— it is the location in which the utterance takes place. In (8b) it is provided by the previous sentence—it is the location of the reporter for the *Times*. In (8c) it is provided sentence internally and is dependent on the domain of the quantifier—for a given choice of sports fan in the domain it is the location of that sports fan.

The variety of readings, then, an implicit argument exhibits depends on the kind of context-dependency involved. Specifically, if it is dependent on the context of utterance, it gets a deictic interpretation, as in (8a), if it is dependent on previous linguistic discourse, it gets a discourse anaphoric reading, as in (8b), and if it is dependent on a quantificational element, it gets a bound variable reading, as in (8c). (8a)–(8c) may have other readings as well—for instance, (8c) can have a reading in which the implicit argument is controlled by the context of utterance— but since our aim here is to delimit the range of possible readings we want to focus on the readings described above. The crucial fact is that the implicit argument *can* be controlled by the context of

---

[8]Fillmore (1986) discusses predicates such as *contribute*, whose implicit theme argument is existential and implicit recipient argument is anaphoric.

[9]The predicate *local* is also vague and, therefore, is context-dependent in another dimension, namely with respect to the degree of precision specifying the range within which something counts as local (Kamp 1975). We will ignore this aspect in what follows. In general, we consider the context-dependency of vague predicates and that of predicates with implicit arguments as distinct phenomena.

utterance, that it *can* be controlled by the previous linguistic discourse, and that it *can* have a bound variable reading when in the scope of a quantificational element.

Certain context-dependent expressions place further restrictions on what types of context can determine their interpretation. For example, *ago* can only pick out its temporal reference relative to the utterance situation, whereas *before* excludes this possibility and can only be controlled by the linguistic context. Thus, in the absence of any prior linguistic discourse, (9a) is felicitous but (9b) is not.

(9)   a. Jill visited Madison two years ago.

        b. # Jill visited Madison two years before.

A somewhat similar contrast exists between *present*, which can only be controlled by the context of utterance, and *current*, which can be controlled either by the context of utterance or the linguistic context, as pointed out by Nunberg (1992).

A theory of context-dependent elements, therefore, must account both for the different types of context-dependency and for the range of readings of elements in each class.

2.3   Anaphora and Implicit Arguments

Partee (1989) distinguished descriptively between three different kinds of contexts: the external context of the utterance, the discourse-level linguistic context and the sentence-internal linguistic context. Taking the bound-variable-like behavior of implicit arguments as central, she argued for the need to provide a sufficiently general notion of context-dependence and to integrate contextual information with "the recursive mechanisms of sentence grammar." We can amplify her point by noting that there are substantive restrictions on what can supply a value for an implicit argument, which are ultimately semantic. One such case was discussed with respect to the contrast between (7a) and (7b). Similarly, (10a) and (10b) differ in that the location of the bar can be dependent on the choice of Italian neighborhood in (10a) but not in (10b) (assuming both indefinite descriptions are within the scope of the adverb of quantification *always*).[10]

(10)   a. When John visits an Italian neighborhood, he always goes to a local bar.

---

[10]The contrasts exemplified by (7) and (10) show that the contextual determination of the interpretation of implicit arguments is not just a matter of pragmatic principles. Compare with Crimmins's (1992:151) claim that "hearers standardly interpret unarticulated constructions [implicit arguments] by relying only on common knowledge and general pragmatic rules, with only the thinnest of semantic rules to go on."

    b. When John goes to a local bar, he always visits an Italian
       neighborhood.

The generalization is one familiar from discussions of anaphoric binding: the element that supplies the interpretation for the implicit argument, the implicit argument's antecedent, must have the predicate
with the implicit argument in its scope. In order to account for this
generalization we must have a theory that would allow us to have a
proper construal of semantic scope.[11] A dynamic framework is best
suited for this task.

    Relying on the semantic theory of Kamp (1981), Partee (1989)
sketches an account in which a context-dependent element indexes to a
context, in this case an accessible DRS. Given the nesting of DRS's, a
context-dependent element may anchor to any superordinate DRS. The
question then is what the requirements are that a context-dependent
element places on the DRS it is indexed to and how they are satisfied. Partee assumes that temporal context-dependent predicates place
the requirement that the DRS they anchor to contain a reference time,
spatial context-dependent predicates place the requirement that it contain a reference location, etc. Although the context-dependent predicate itself does not have as an argument a variable corresponding to a
reference time or location, the DRS it anchors to must.[12]

    More generally, Partee, seconded by Nunberg (1992), claims that
a dependent element—be it a pronoun or the implicit argument of a
relational predicate—has associated with it the specifications in (11),
which account for its interpretation.

(11)   a. what kinds of context it can anchor to
       b. requirements on the context for the element to be felicitous
       c. meaning

According to Partee and Nunberg, (11a) is needed to distinguish between different types of context-dependent elements, such as those that
can only be controlled by the context of utterance (e.g., pure indexicals,
implicit argument of *ago, present*), those that can be controlled by any
one of the three kinds of contexts (e.g., pronouns, implicit argument of
local) and those that exclude control by the context of utterance (e.g.,

---

[11]Syntactic constraints can enter the picture at this point since they play a role in
determining semantic scope.

[12]This is assuming that reference times, locations, etc., are treated as a special type
of discourse referents, as in the treatment of temporal anaphora in Kamp & Rohrer
(1983), Partee (1984), Hinrichs (1986). Alternatively, reference times, locations,
etc. can be construed as parameters of evaluation for a context. The issues we raise
do not hinge on this choice and can be reformulated appropriately.

reflexives, *before*).[13] The challenge for a unified analysis of context-dependency is to characterize the three kinds of contexts in a uniform way without loosing the distinction between them.

Kamp (1981) and Heim (1982) unified the deictic use of pronouns with all their other uses by assuming that the top-level discourse context subsumes the external context of the utterance. Information about individuals salient in the context of utterance and the "guise" (in the sense of Lewis (1979)) under which they are salient is accommodated in the top-level context.[14] As Heim (1982) puts it, deictic reference is mediated by the file and the context of utterance supplies "not simply individuals but 'individuals in guises'" (p. 318). In taking deictic pronouns to be only context-sensitive, Kamp and Heim departed from previous analyses of deictic uses of pronouns, which took them to be both context-sensitive and directly referential. These analyses are based on Kaplan's (1989a) fundamental distinction between context and circumstances of evaluation. According to such analyses, the denotation of a deictic pronoun might vary from context of utterance to context of utterance, but once the context of utterance is fixed, then its denotation does not vary from one circumstance of evaluation to another. Having deictic reference mediated by a file results in a radically different theory, since then deictic reference does not involve direct reference and, therefore, the denotation of a deictic pronoun can vary from one circumstance of evaluation to another, relative to the same context of utterance.[15]

Partee differs from Kamp and Heim in keeping the external context separate; in the representations she gives, it is the outermost context and every other context is nested within it.[16] The external context encodes information about the utterance situation, such as the speaker, the time and location of the utterance.

2.4 Indexicality and Implicit Arguments

The separation of the external context from the top-level discourse context is necessitated by the semantic properties of pure indexicals. Pure indexicals such as *I, here, now* anchor to the external context and are directly referential. They cannot be given an account parallel to that of deictic pronouns in Kamp (1981) and Heim (1982) for the rea-

---

[13] Plural context-dependent elements may combine the different types of context-dependency. See the discussion on the first person plural pronoun *we* in Partee (1989) and Nunberg (1993).

[14] A similar proposal is made by Karttunen (1969), whereby contextual salience is sufficient for the introduction of a discourse referent.

[15] For an indication of the consequences of this move see Heim's discussion on presupposed coreference and deictic pronouns (Ch. III, Sec. 2.3).

[16] See Kamp (1990) for an explicit treatment of indexicality within DRT.

sons that motivated direct reference theories in the first place (Kaplan 1979, 1989a, 1989b, Perry 1977 *inter alia*). The descriptive conditions associated with their meaning are responsible for picking out the right individual in a given context of utterance but do not enter the content of what is said. As Kaplan (1989a, 500) puts it, "the descriptive meaning of a pure indexical determines the referent of the indexical with respect to a context of use but is either inapplicable or irrelevant to determining a referent with respect to a circumstance of evaluation."

As a consequence, (12a) is not synonymous with (12b): the proposition expressed by (12b) relative to any context of utterance is necessarily true, while that expressed by (12a) is not.

(12)   a. I am the speaker (of this utterance).
        b. The speaker (of this utterance) is the speaker (of this utterance).

Following Nunberg (1992, 1993), we will use the term *indicativeness* to characterize the property of certain linguistic elements (such as pure indexicals) that the conditions determining their reference do not enter into their content. We intend this only as a descriptive term; those elements that are indicative must be so as a result of their meaning specification.

Like indexicals, implicit arguments can be controlled by the context of utterance. Can they, like indexicals, be indicative? Nunberg (1992) claims that control by the context of utterance and indicativeness do not go hand in hand and uses, in fact, implicit arguments as a case in point. We will argue that implicit arguments have indexical readings and in that case they are indicative.

In the following section we show that implicit arguments pattern with definite descriptions on their anaphoric reading and in section 4 we argue that they behave like true indexicals on their deictic reading.

3   Implicit Arguments as Implicit Descriptions

3.1   Contexts for Implicit Arguments

What information should the context provide for the interpretation of implicit arguments? In the examples considered so far where the implicit argument is not controlled by the utterance situation, there is an overt linguistic element that can be identified as the antecedent of the implicit argument. Within a theory that allows for dynamic binding, one can analyze the antecedent as directly binding the implicit argument. Furthermore, Partee (1989) assumed that a subset of open-class context-dependent expressions exploit those aspects of contexts that are always present and always unique, such as reference time, reference location, and point of view. The reference time, reference location and point of view of the external context can be assumed to be, respectively, the time, place and speaker of the utterance. Putting the two assumptions together, we can formulate the hypothesis that an implicit argument is bound either by an overt accessible antecedent or by some implicit element which is always present in a context and unique relative to that context.[17] However, these cases do not exhaust the range of possibilities: there need not be an *overt* antecedent,[18] even when the implicit argument is not controlled by some element always present in a context.

As we will show below, contexts for predicates with implicit arguments are those that entail, or can be extended (via accommodation) so as to entail, the existence of an entity that satisfies the selectional restrictions imposed by the predicate for that argument. In that respect implicit arguments pattern with definite descriptions rather than pronouns in allowing for associative anaphoric readings. Saying, then, that implicit arguments function like pronouns in terms of the context-dependency they exhibit is true in one respect but not in another: implicit arguments are like pronouns in their capacity to anchor to any kind of context; however, they are unlike pronouns and more like definite descriptions in not demanding an overt antecedent.

The sentences in (13) illustrate the way in which implicit arguments pattern with definite descriptions. Even without an overt antecedent denoting a bet, the sentences (13a) and (13b) share a reading on which every man won the wager he made on the outcome of the Superbowl. The pronoun in (13c) lacks this reading. Clearly, (13a) does not constitute a case where the implicit argument is controlled by some fixed aspect of context, since contexts in general do not contain information about the existence of bets.

(13)   a. Every man who bet on the Superbowl won.
        b. Every man who bet on the Superbowl won the bet.
        c. Every man who bet on the Superbowl won it.

Given the restriction of the quantifier, for every man in the domain of quantification there is entailed to be a bet and that is sufficient for the interpretation of the implicit argument of *win*. It is exactly this

---

[17]Partee surmised that these two options correspond to different types of context-dependency and that therefore the classes of context-dependent elements are distinguished along these lines—*third-person-like dependents* and *first-person-like dependents*, as she calls them. The analysis we develop accepts that there is a division but groups the class of implicit arguments discussed so far (except for those that are always indexical) in the first rather than the second category. See the discussion in section 4.2.

[18]Partee (1989) made a similar observation but she did not draw from it the conclusions we are drawing.

entailment that is responsible for the felicity of the dependent definite *the bet* in (13b) as well.

Thus, implicit arguments have a wider range of readings than pronouns. Implicit arguments also have a wider range of readings than elements which are strictly dependent on reference parameters. Temporal expressions with implicit arguments, for example, need not be anaphoric on the reference time of any accessible context. For instance, (14a) has a reading which is equivalent to that of (14b) although there is no reason to assume that the context of the restriction of the quantifier contains a reference time corresponding to the time of the escape.[19] Similarly, (14c) has a reading which is equivalent to that of (14d) even if nothing in the previous discourse has triggered the introduction of a reference time corresponding to the time of the baking of the cake.

(14)   a. Every fugitive was caught within a month.
       b. Every fugitive was caught within a month of the time he escaped authority control.
       c. The cake tasted better after two days.
       d. The cake tasted better two days after the time it was baked.

Being a fugitive entails having escaped authority control at some point and this entailment is sufficient to provide the information for the implicit argument of *within a month*, just as it is to provide the information for the felicity of the dependent definite *the time he escaped authority control*. As for (14c), from the fact that something is a cake we can infer that it was baked; a context with information about a cake then can be extended to a context with information about the time of the baking of the cake.

A piece of evidence that there is no reference time corresponding to the time of the escape in the restriction of the quantifier in (14a), or a reference time corresponding to the baking of the cake in (14c)[20] is the contrast between the implicit argument predicates in (14) and those that can only be anaphoric on a reference time, such as the adverbs *afterwards, beforehand, thereafter*. While (15a) has a reading equivalent to that of (15b), (15c) does not.[21] Similarly, for (15f) the implicit

---

[19]Even if the nominal predicate *fugitive* is assumed to have a temporal argument (e.g., Enç 1986), the value of that temporal argument is the interval during which a given individual was a fugitive, not the time of his escape.

[20]This is assuming that nothing in the previous discourse has introduced such a reference time.

[21]Adverbs that are anaphoric on a reference time have the bound variable reading only if there is quantification over reference times, as is the case with *when*-clauses:

    Whenever a prisoner escaped, he became a respectable citizen some time afterwards.

argument of *afterwards* cannot be taken to be the time of the baking of the cake unless that time is already introduced in the discourse.

(15)   a. Every fugitive became a respectable citizen after some time.
       b. Every fugitive became a respectable citizen after some time from the time he escaped authority control.
       c. Every fugitive became a respectable citizen some time afterwards.
       d. The cake tasted good even after a long time.
       e. The cake tasted good even after a long time from the time it was baked.
       f. The cake tasted good even a long time afterwards.

For (15c) and (15f) there must be some salient time in the discourse relative to which *afterwards* is evaluated. Thus, the analysis that Partee (1989) proposed for the entire class of context-dependent predicates with anaphoric readings seems to fit best a restricted subclass of them.

An implicit argument can also be dependent on a context which is itself implicit, i.e., does not correspond to any overt linguistic material. For example, (16a) has a reading equivalent to that of (16b) and one equivalent to that of (16c). The implicit argument of *within a day*, then, can be construed so as to get the available linguistic element as its antecedent and so as to be equivalent with a definite description taking narrow scope with respect to the modal.

(16)   a. He might leave today. He should return within a day.
       b. He might leave today. He should return tomorrow.
       c. He might leave today. He should return within a day of the day he leaves.

Although the modal *should* has no overt restriction, because of the previous sentence in the discourse it is taken to quantify over those possible worlds in which he leaves (today or at some time or other). This is a case of modal subordination (Roberts 1989) and the interpretation of the implicit argument depends on the way the restriction of the modal quantifier is construed.[22] *Today*, although a potential antecedent, does not have to bind the implicit argument nor does it have to determine the implicit restriction. The time of leaving is dependent upon the choice of modal base for the modal *should*, which does not have to be chosen in such a way that the time of leaving is today.

3.2   Our Proposal

In our analysis of implicit arguments, we will reconstruct the requirement that an implicit argument places on a context (Partee's (11b)

---

[22] As we will discuss below, the implicit restriction does not have to be an explicit piece of some representation.

above) as well as the specifications of what kinds of contexts are acceptable anchors (Partee's (11a) above) as felicity conditions. Following recent work in dynamic semantics (Heim 1982, Groenendijk & Stokhof 1991, Dekker 1993), we construe contexts as information states and take the meaning of a sentence to be specified in terms of how it updates a given information state. Meanings, in general, are characterized as possibly partial functions from information states to information states. Following Heim (1982, 1983) and Beaver (1993), we take felicity conditions to be requirements on an information state in order for the meaning function to be defined on that information state. An information state is specified as follows:

An information state S is a set of pairs of worlds and assignments such that all the assignments have the same domain, which we call V_S. We call the set of worlds determined by S W_S, and the set of assignments determined by S F_S.

We base our analysis of implicit arguments on Heim's (1982) analysis of definite NPs, according to which their interpretation depends on the felicity conditions they impose on input information states. Definite descriptions require familiarity of their corresponding variable and their descriptive content. (17) exemplifies the analysis of standard definite descriptions.

(17) S ⟦ the man$_x$ plays duets ⟧ =
      $\{\langle w, f\rangle \in S \mid f(x) \in w(\text{play-duets})\}$
      iff $x \in V_S$ and $\forall \langle w, f\rangle \in S$: $f(x) \in w(\text{man})$.
      Else undefined.

We analyze the implicit arguments of relational predicates on a par with definite descriptions. Implicit arguments are associated with familiarity conditions for their corresponding variable and their descriptive content. The descriptive conditions associated with an implicit argument are, minimally, the sortal properties required by the predicate the implicit argument is an argument of.

(18b) illustrates our account of the implicit argument of *local*.[23] The only felicity condition placed by the predicate *local* is that the filler of its second argument be a location (essentially just the selectional

---

[23] If the indexing to contexts approach is spelled out in a particular way that recognizes *local* to be a relational predicate, then as far as we can see, it is not substantially different from the proposal put forth here. For instance, some world-assignment pair $\langle w, f\rangle$ would satisfy the condition $local_C(x)$ iff $\langle f(x), f(y)\rangle$ $\in w(\text{local})$, where y is some discourse marker in C such that if $\langle w', g\rangle$ satisfies C then $g(y) \in w'(\text{location})$. This formulation is appropriate for cases in which the implicit argument exploits a feature of context that is always present. For other cases, we would have to distinguish between licit and illicit indexings to a context according to whether that context contains the information necessary for the interpretation of the implicit argument.

restriction on that argument). What reading the implicit argument gets depends on what information S contains. (18d) illustrates the way the discourse anaphoric reading comes about, assuming S is the information state resulting from updating with the first sentence of (18c).[24]

(18)   a.   A local$_z$ bar$_y$ was selling cheap beer.
       b.   S ⟦ a local$_z$ bar$_y$ was selling cheap beer ⟧ =
            $\{\langle w, f \rangle \mid \exists \langle w, g \rangle \in S\!: g <_y f, \ f(y) \in w(\text{bar})$
            $\langle f(y), f(z) \rangle \in w(\text{local}), \ f(y) \in w(\text{sell-cheap-beer})\}$
            iff $y \notin V_S$ and $z \in V_S$ and $\forall \langle w, f \rangle \in S\!: f(z) \in w(\text{location})$.
            Else undefined.
       c.   A reporter$_x$ for the Times got seriously drunk.
            A local$_z$ bar$_y$ was selling cheap beer.
       d.   S ⟦ a local$_z$ bar$_y$ was selling cheap beer ⟧ =
            $\{\langle w, f \rangle \mid \exists \langle w, g \rangle \in S'\!: g <_y f, \ f(y) \in w(\text{bar})$,
            $\langle f(y), f(z) \rangle \in w(\text{local}), \ f(y) \in w(\text{sell-cheap-beer})\}$
            iff $y \notin V_{S'}$ and $z \in V_{S'}$ and $\forall \langle w, f \rangle \in S'\!: f(z) \in w(\text{location})$,
            where $S' = \{\langle w, f \rangle \mid \ \exists \langle w, f' \rangle \in S\!: f' <_z f$ and
            $\langle f(z), f(x) \rangle \in w(\text{location-of})\}$.
            Else undefined.

The difference between S and S′ is accommodated information. In order for the felicity conditions of *local* to be satisfied, S′ accommodates a new variable, z, assigned to a location and related to the variable x in S, assigned to a reporter. This is why in the reading represented for (18c), the bar is understood to be local to the reporter for the *Times* mentioned in the previous sentence.

The accommodation of the necessary information in order to satisfy an implicit argument's felicity conditions is analoguous to the accommodation brought about in order to satisfy the felicity conditions of a definite NP which is not directly anaphorically related to anything mentioned in the discourse thus far:

(19)   The Porsche lurched to a stop. The engine was smoking.

The engine in (19) is understood as the engine of the Porsche. In accommodating the information for the felicity of the definite we satisfy the felicity conditions of the definite by connecting the newly introduced entity to some already mentioned entity in the discourse (Heim 1982).

---

[24]The meaning specifications we provide throughout the paper make use of three relations between assignment functions, defined as follows:

$g \leq f$ iff $\text{Dom}(g) \subseteq \text{Dom}(f)$ and $\forall v \in \text{Dom}(g)\!: f(v) = g(v)$
$g < f$ iff $\text{Dom}(g) \subset \text{Dom}(f)$ and $\forall v \in \text{Dom}(g)\!: f(v) = g(v)$
$g <_y f$ iff $g < f$ and $\text{Dom}(f) = \text{Dom}(g) \cup \{y\}$

Accommodation is necessary in (19) to capture the dependency of the definite *the engine* on the NP *the Porsche* as well as to get the right truth conditions for the NP *the engine*.

Following Heim, the kind of accommodation illustrated in (18d) and (19) basically implements the following strategy:

Ordinary Accommodation
The information necessary to satisfy the familiarity condition of a definite NP may be accommodated to a state S, yielding a new state S′, by relating the definite's discourse marker through some relation to a discourse marker in the domain of S.

If the definite's discourse marker is x, the relation is R, and the discourse marker to be related to x is y, then for an input state S, we can define a new state by way of a 4-place function on S, R, x, and y:

$$\text{Accom}(S, R, x, y) = \{\langle w, f\rangle \mid \exists\langle w, g\rangle \in S: g <_x f \text{ and } \langle f(x), f(y)\rangle \in w(R)\}$$

There are various kinds of conditions one might put on Accom to refine the theory of accommodation, such as that the world set of S and S′ be the same (the accommodated entity was entailed to exist), and that R be functional, but these would lead us far afield into tangled issues in the semantics of definites. For now, we assume an underconstrained theory of Accom, which doubtlessly allows too much to be accommodated.

Using Accom, (18d) can be reformulated as in (20).

(20) S ⟦ a local$_z$ bar$_y$ was selling cheap beer ⟧ =
$\{\langle w, f\rangle \mid \exists\langle w, g\rangle \in \text{Accom}(S, \text{location-of}, z, x): g <_y f,$
$f(y) \in w(\text{bar}), \langle f(y), f(z)\rangle \in w(\text{local}),$
$f(y) \in w(\text{sell-cheap-beer})\}$
iff $y \notin V_S$ and $\forall\langle w, f\rangle \in \text{Accom}(S, \text{location-of}, z, x):$
$f(z) \in w(\text{location}).$
Else undefined.

And if the first argument of the relation location-of is guaranteed by a meaning postulate to be a location, this becomes simply:

(21) S ⟦ a local$_z$ bar$_y$ was selling cheap beer ⟧ =
$\{\langle w, f\rangle \mid \exists\langle w, g\rangle \in \text{Accom}(S, \text{location-of}, z, x): g <_y f,$
$f(y) \in w(\text{bar}), \langle f(y), f(z)\rangle \in w(\text{local}),$
$f(y) \in w(\text{sell-cheap-beer})\}$
iff $y \notin V_S$.
Else undefined.

An implicit argument can exhibit scopal effects through ordinary accommodation. Its scope, as it were, is determined by the information state that entails its presuppositions. When the implicit argument is within a quantificational context, then its presuppositional conditions can be satisfied by the auxiliary information states that come about in the calculation of the information update brought about by a quantifier. In that case, the implicit argument has the bound-variable-like reading, or more appropriately in our terms, a dependent reading (as with dependent definite descriptions). This is illustrated in (22).

(22) a. Every man$_x$ who bet on the Superbowl won$_z$.
   b. S $[\![$ every man$_x$ who bet on the Superbowl won$_z$ $]\!]$ =
   $\{\langle w, f\rangle \in S \mid \forall h: f <_x h$ [if $\langle w, h\rangle \in S'$, then $\exists g: h \leq g$,
   $\langle w, g\rangle \in$ Accom$(S',$ bet-of, $z,$ $x)[\![$won$(x, z)]\!]$]$\}$
   iff $x \notin V_S$ and $\forall\langle w, h\rangle \in$ Accom$(S',$ bet-of, $z,$ $x)$:
   $h(z) \in w($contest$)$
   where $S' = \{\langle w, f\rangle \in S \mid \exists g: f <_x g,$ $g(x) \in w($man$),$
   $g(x) \in w($bet-on-the-Superbowl$)\}$
   Else undefined.

The narrow scope readings with respect to a modal, as in (16a), arise when the information state construed on the basis of the modal base of the modal is compatible with the information necessary to satisfy the felicity conditions of the implicit argument and this information can thus be accommodated relative to that information state. The way the narrow scope reading of the implicit argument of (16a) arises is shown in (23). The modals are taken to have an epistemic modal base; Acc(w) is a set of epistemically accessible worlds from w.

(23) a. He$_x$ might leave today. He$_x$ should return within$_t$ a day.
   b. S $[\![$ he$_x$ might leave today $]\!]$ = S' =
   $\{\langle w, f\rangle \in S \mid \{\langle w', f\rangle \mid w' \in$ Acc$(w)\}[\![$he$_x$leaves today$]\!] \neq \emptyset\}$

   c. S* $[\![$ he$_x$ should return within$_t$ a day $]\!]$ =
   $\{\langle w, f\rangle \in$ S* $\mid S_{acc,w}[\![$he$_x$returns within$_t$a day$]\!] = S'_{acc,w}\}$
   where S* = $\{\langle w, g\rangle \mid \exists\langle w, h\rangle \in$ S': $h <_t g,$ $\forall w' \in$ Acc(w):
   $g(t) \in w'($time$),$ $\langle g(x), g(t)\rangle \in w'($leave$)\},$
   $S_{acc,w} = \{\langle w', g\rangle \mid w' \in$ Acc$(w),$ $f <_t g,$ where
   $\langle w, f\rangle \in$ S*$\}$ and
   $S'_{acc,w}$ is such that $W_{S_{acc,w}} = W_{S'_{acc,w}}$ and $V_{S_{acc,w}} \subseteq V_{S'_{acc,w}}.$

S* is an information state such that its world set is comprised of worlds in which he leaves at some time and it has a discourse marker for the time of the leaving. S* then satisfies the felicity conditions of the implicit argument of *within* and can thus be updated with *he$_x$ returns within$_t$ a day*.

Having an overt variable corresponding to the implicit argument, we do not require accommodation at some representational level since we specify the presuppositions and the anchoring conditions of the implicit argument in terms of definedness conditions on an information state. The indexing to contexts approach, on the other hand, cannot sidestep the need for representational accommodation.

Assuming that contentful context-dependent elements correspond to relational predicates with an implicit argument does not commit us to the presence of a phonologically null argument at the syntactic level. It also does not commit us to there being a one-to-one syntactic or semantic correspondence between an expression with an implicit argument and one with an overt argument. There are clearly syntactic differences between the two: syntactically present arguments require case marking and they might require certain prepositions in addition (*a local bar, a bar local to the neighborhood*); predicates with implicit arguments may have a different syntactic distribution than predicates whose arguments are syntactically expressed. For instance, predicates that are nominal modifiers may appear prenominally if they have no syntactic complements and postnominally if they do (*a local bar, a bar local to the neighborhood*). As pointed out in section 2.1, the selectional restrictions of a predicate with an overt argument and the corresponding one with an implicit argument may also be different.

Finally, if an overt argument is purely anaphoric, then it must have an overt antecedent. As we have seen, no such requirement exists for an implicit argument. This we believe is the right perspective to have on such differences as illustrated by the contrast in (24), discussed by Partee (1989).

(24)   a. In all my travels, whenever I have called for a doctor, one has arrived within an hour.
        b. # In all my travels, whenever I have called for a doctor, one has arrived there within an hour.

There is anaphoric and like personal pronouns it requires an overt antecedent. Adverbial elements that require an overt antecedent also share with personal pronouns the property of requiring an overt antecedent even when the antecedent does not semantically bind them. Informally, we can say that all semantically anaphoric elements can have E-type readings, as for example in (25). In that respect then anaphoric elements requiring an explicit antecedent and those that do not, such as definite descriptions and implicit arguments, are similar.

(25)   a. He wants to move to a big city. But there must be a good beach not far away.
        b. He wants to move to a big city. But there must be a good beach not far away from there.

c. He wants to move to a big city. But there must be a good beach not far away from the center.

How to formulate a theory that can capture this notion of antecedent-hood remains an open problem and we will not address it here.[25]

4 Indexicality and Implicit Arguments

4.1 Unifying the Analysis

In this section we provide an account of indexicals and of the indexical readings of implicit arguments. Thus far, our account of implicit arguments treats them like definites and invokes the device of accommodation to handle a number of crucial cases. This suggests a natural strategy for handling the indexical readings of implicit arguments, such as that of (8a), repeated here.

(26)   A local bar was selling cheap beer.

To handle the reading on which locality is relative to the speaker, why not simply accommodate one discourse marker for the speaker and another for the speaker's location?

One immediate obstacle to this account is that accommodation as we have defined it requires an object already in the discourse and a relation by which to accommodate, and yet (26) seems to retain the indexical reading in almost any context. There is also something a little unnatural about an account which can introduce speakers into discourses only by way of something prior and more familiar. Invoking accommodation for the indexical readings, then, would probably require enriching our theory of accommodation, something we should do only with great care.

But there is a much more important reason for not handling the indexical readings by appealing to more accommodation: in their indexical readings, implicit arguments do not behave as definites do, but rather as indexicals do.

A property of indexicals that plays a central role in motivating the account of Kaplan (1989a) is their *referentiality*. For example, if the second author of this paper utters (27a), he expresses the proposition that it ought to be the case that Mark Gawron is a woman. On the other hand, if Mark Gawron utters (27b), he may express that same deontic judgement, but he can also express another, more sweeping judgement, that it ought to be the case that whoever makes this utterance is a woman.

(27)   a. I ought to be a woman.
         b. The speaker of this utterance ought to be a woman.

---

[25]It is discussed in some length by Heim (1990) and Chierchia (1992).

Thus, the definite in (27b) can be captured by the deontic operator, the indexical in (27a) cannot. Kaplan argued that no operator can capture an indexical in its scope.

More generally, the referentiality property of indexicals is this: relative to a given context of utterance an indexical refers to a particular individual—the unique individual who in that context satisfies the descriptive conditions associated with the meaning of the indexical—and reference to that individual persists, regardless of the circumstances of evaluation relative to which other parts of the sentence in which the indexical occurs are evaluated, and regardless of what that indexical may be referring to relative to other contexts of utterance. Kaplan captures the referentiality property of indexicals by assuming that indexicals are directly referential and by postulating a condition ruling out operators that operate on characters.

The problem Kaplan's analysis faces from the point of view of a unified account of context-dependency is that it is forced to claiming ambiguity for any expression that can have both an indexical and a non-indexical reading.

Implicit arguments that anchor to the external context exhibit the referentiality property of pure indexicals. Consider the sentences in (28), all uttered in a context in which the speaker is arriving at an outdoor location and is asked to evaluate the environs for their aesthetic merits:

(28)   a. There could have been a river near here.
     b. There could have been a river near the location of this utterance.
     c. There could have been a river nearby.

Examples (28a) and (28b) show the expected contrast between indexicals and definite descriptions. If (28b) is in fact uttered in Palo Alto, a riverless city, then it can be made true by counterfactual alternatives in which that sentence is uttered at a place that does have a river, say, in the woods of the Sierra foothills. If, on the other hand, (28a) is uttered in Palo Alto, it remains stubbornly anchored to Palo Alto. The only counterfactual alternatives that can make it true involve geological or climactic changes that steer a river through Palo Alto. Thus, the indexical remains referential, its reference constraining the set of alternatives we consider. Example (28c) has the reading of (28a), but not of (28b). In a context in which the implicit argument is constrained to be the location of the utterance, the implicit argument patterns with the indexical. If implicit arguments can have what can be described as "narrow scope readings" with respect to modal operators (as in (16a) and (25a)), why is that not the case in (28c)?

The challenge for us is to integrate the analysis provided in section 3 within a proper theory of indexicality. Following Kaplan, we will pursue an account not by providing a mechanism that always assigns indexicals wide scope, but by making them obligatorily referential.

4.2 Sketch of an Account of Indexicals

In Kaplan's theory, meanings (characters in his terms) are functions from contexts to contents. The meaning of a sentence, for example, is a function from contexts to sets of circumstances of evaluation. The truth-conditional content of a sentence with respect to a context c is the set of all circumstances of evaluation with respect to which the sentence is true relative to c. In a dynamic theory that takes contexts into account, meanings should be functions from contexts to functions from information states to information states:

Revising Sentence Meaning
Let Σ be the set of information-states. Let C be the set of contexts (to be explained below). Then sentence meanings are elements of

$$(\Sigma^\Sigma)^C.$$

For a sentence ϕ we will symbolize the result of applying $[\![\phi]\!]$ to a context c as $[\![\phi]\!]^c$.

A context c is an information state such that F_c contains only a single assignment function f_c and W_c contains only a single world w_c, the world in which a given utterance takes place. Every expression α,[26] context c and state S that is an argument of $[\![\alpha]\!]^c$ must satisfy the Utterance Condition and the Consistency Condition, given below.

- **Utterance Condition**
  There is a **u** which is an utterance token of α in w_c, that is,

  $$\langle \mathbf{u}, \alpha \rangle \in w_c(\text{utterance-token-of})$$

  and there are **u**, **s**, **l**, and **t**, with **u**, **s**, **l**, **t** all distinct, such that

  $$\langle \mathbf{s}, \mathbf{u}, \mathbf{l}, \mathbf{t} \rangle \in w_c(\text{speaking-at}).$$

  There must also be three variables x, y, z, such that

  $$\{x, y, z\} \subseteq V_c \text{ and}$$
  $$f_c(x) = \mathbf{s}, \ f_c(y) = \mathbf{l}, \ f_c(z) = \mathbf{t}.$$

We refer to the unique **s**, **l**, **t** selected by the f_c of some context c as **ego**$_c$, **here**$_c$, and **now**$_c$, respectively. We refer to them collectively as *contextual roles*. Contextual roles are features of an utterance situation and any context must contain information about them, if it is to satisfy the Utterance Condition, regardless of whether a language has an

---

[26]We assume the expressions of the language have been disambiguated; that is, they are expressions of the sort to which we assign meanings.

indexical expression that picks one of them as its referent.[27] A context that satisfies the Utterance Condition is such that it characterizes an utterance situation: for a world to be in the world set of such a context an act of utterance must be taking place in it.[28]

- **Consistency Condition**
  S and c must be consistent.
  An information state S and a context c are consistent if and only if

$$\forall f \in F_{\text{S}}: f_c \leq f.$$

The Consistency Condition requires that all the assignments made by a context be preserved by the assignment functions of an information state.

Pure indexicals, like *I, here, now*, as well as expressions with indexical implicit arguments, such as *ago, present*, have a variable associated with them, like other NPs and implicit argument predicates. That variable must be assigned to the right individual corresponding to a given contextual role. Indexical elements thus place requirements on information states in order for the meaning of the sentence in which they occur to be defined and are therefore associated with certain felicity conditions. Indexicals (or expressions containing indexicals) are felicitous only relative to information states assigning their corresponding variable to the right individual supplied by the context.[29] For any context c that satisfies the Utterance Condition and any information state S that is consistent with it, it will be the case that every assignment function in F_{S} assigns some variable to $\mathbf{ego}_c$, $\mathbf{here}_c$ and $\mathbf{now}_c$.

The meaning of a sentence with *I* can then be specified as in (29).[30]

---

[27] Given our account of the indexical readings of implicit arguments below, it follows that a language that does not have an indexical corresponding to English *here* can still have implicit arguments with an indexical reading. Such implicit arguments would select $\mathbf{here}_c$ as their referent relative to context c.

[28] For Kaplan, a context must include an agent but need not include an utterance. He characterizes occurrences of a sentence in a context, rather than utterances of a sentence in a context. Here we are following the lead of Barwise & Perry (1983) in trying to derive some important properties of indexicals from their connection to utterances in utterance situations.

[29] We are assuming free indexing so an indexical might bear any index whatsoever; illicit indexings are ruled out by the felicity conditions associated with the indexical. An occurrence of an indexical such as I_r might be infelicitous relative to an information state S and a context c even though c satisfies the Utterance Condition and therefore contains an $\mathbf{ego}_c$. This will be the case when S entails that $f(r)$ is distinct from $\mathbf{ego}_c$.

[30] On this proposal, the meaning of *I*, and of indexicals in general, is such that it yields a constant character: for instance, for any c and c', S $[\![$ I_r play duets $]\!]^c$ and S $[\![$ I_r play duets $]\!]^{c'}$ are equal if both are defined.

(29)   S $[\![$ I$_r$ play duets $]\!]^c =$
       $\{\langle w, f \rangle \mid \langle w, f \rangle \in$ S and $f(r) \in w(\text{play-duets})\}$
       iff $\forall f \in F_S$: $f(r) = \textbf{ego}_c$.
       Else undefined.

Relative to a given context of utterance then, an indexical, in effect, refers to a particular individual, namely the unique individual who in that context fulfills the contextual role required by the felicity condition associated with the indexical. Relative to that context of utterance the indexical would refer to that individual with respect to any information state that is consistent with the context, regardless of whether the information entailed by the information state about that individual is compatible with the information entailed by the context.[31]

We turn now to accounting for the possibility of an indexical reading for implicit argument predicates like *local* or *nearby*. According to the account of implicit arguments presented in section 3 and our new treatment of context, the implicit argument of such predicates will now *necessarily* be allowed an indexical interpretation in any context. The indexical interpretation is allowed because in any context c, the location role of *local* can be filled with $\textbf{here}_c$. Consider (8a) and (28c). The Consistency Condition guarantees that any information state for which the meaning function relative to a context c is defined will be such that all its assignment functions assign some variable to $\textbf{here}_c$. Thus, any interpretation using $\textbf{here}_c$ will satisfy the presuppositional requirements of *local* or *nearby*. In general, anchoring to the context of utterance for an implicit argument means that its corresponding variable is assigned to one of the contextual roles.

On our account, the indexical interpretation of an implicit argument functions as a kind of default. The utterance context is always present and the information it provides always available via the information state. So in the absence of any linguistic context generating entailments to fill the implicit argument role the utterance context will serve. It is perhaps no accident, then, that the vocabulary of implicit arguments is replete with words connected with space and time, when these are two of the principal features of the utterance context.

More generally, the kinds of contexts a dependent element can anchor to (as described in section 2) depends on the specificity of its felicity conditions and need not be stated independently. The elements that are descriptively characterized as anchoring to the context of utterance are those whose felicity conditions require that they be assigned to some contextual role. The elements that are descriptively characterized

---

[31]This implies, for example, that *I* picks out the **ego** determined by a context c even relative to information states that contain the information that the individual satisfying the contextual role of **ego** in c is not uttering anything.

as anchoring to any context whatsoever are those whose felicity conditions require that they be assigned to some familiar (in the technical sense) entity.[32]

4.3 Operators and Implicit Arguments

We turn now to examining how the behavior of implicit arguments under operators differs from that of definite descriptions. We have claimed that the implicit argument in (28c) has an indexical interpretation, i.e., the implicit argument is indicative and referential. In the right context, however, the implicit argument can be interpreted as non-referential. Consider (30):

(30)   a. If I had uttered this sentence in a forest, there could have been a river nearby.
        b. If I had uttered this sentence in a forest, there could have been a river near the location of this utterance.

Example (30a) has a reading very close to the alternative-utterance-locations reading of (30b). Consider once again example (28), reproduced here as (31):

(31)   a. There could have been a river near here.
        b. There could have been a river near the location of this utterance.
        c. There could have been a river nearby.

Recall that in contrast to (31b), (31c), in a context in which the implicit argument is constrained to be the location of the utterance, lacks the reading on which alternative utterance locations are considered (a reading available for 31b). One available reading is an indexical reading like that of (31a). Example (30) makes it clear that the absence of the alternative-locations reading has nothing to do with the *meaning* of (31c), but with how it is evaluated in the absence of any contextual help.

In fact, given some natural assumptions about the possibility of accommodation, the contrast in the available readings for (31b) and (31c) is predicted by our account. Let us first see how the non-referential reading of (31b) is possible.

In a context providing no discourse marker for the utterance and its location, the definite in (31b) can only be licensed by ordinary accommodation; as discussed in Section 3, ordinary accommodation involves

---

[32]Elements that are descriptively characterized as anchoring to any context but the context of utterance might not form a homogeneous class. The properties of reflexives should follow from their requirement for syntactic binding, rather than any special felicity conditions. Elements like *before*, on the other hand, can be associated with an extra felicity conditions such that their anchoring to the context of utterance is excluded.

relating a new discourse marker to some marker already in the domain of the discourse.

Now let us consider the definite in (31b). On the relevant reading, the most natural way to accommodate the definite *the location of this utterance* is given by the content of the noun phrase itself. That is, if the definite's marker is l and the marker for the utterance is u, the definite is accommodated by way of the relation:

(32)   location-of(l,u)

But there is no reason to expect a neutral context to supply the discourse marker u for the utterance. Thus, this accommodation is only possible if there is a prior accommodation, called for by the presence of the demonstrative *this utterance*.

This is not the place to sketch out a full account of demonstratives. The key point is that a felicitous utterance of a demonstrative requires that its denotation be in the discourse. Thus, having a discourse marker u for the utterance in a neutral context is possible because the demonstrative *this utterance* forces one to be in the discourse. Let us assume for concreteness that demonstratives are definites accommodated by relating the demonstrative discourse marker to the demonstration that introduces the definite.

For the sake of illustration, let us assume an account of counterfactuals along the lines of Stalnaker (1968) and Lewis (1973). We assume a revision relation Revises, based on a world similarity relation, such that

$$\text{Revises}(w_0, w_1, W)$$

is true if and only world $w_0 \in W$, W a set of worlds, and there is no element of W that is more similar to w_1 than w_0. Note that in requiring w_0 to be in W we exclude the case where W is empty. This means that when W is a necessarily false proposition (the empty set), the Revises relation never holds.

To capture the semantics of counterfactual *could*, we also need a function that takes us from the meaning of the antecedent of a counterfactual evaluated against a context c, $[\![\phi]\!]^c$, to the proposition (set of worlds) that meaning expresses in that context. We call this function $\mathcal{W}$:

$$\mathcal{W}([\![\phi]\!]^c) = \{w \mid \text{There is a state S, } w \in W_S, \text{ and a state S}',$$
$$\text{S}' = \text{S}[\![\phi]\!]^c, \text{ and } w \in W_{S'}\}$$

The set $\mathcal{W}([\![\phi]\!]^c)$ consists of worlds w such that w is both in the world set of some state S and in the world set of S', the result of updating S with ϕ. Thus, $\mathcal{W}([\![\phi]\!]^c)$ is just the set of worlds that makes ϕ true relative to c.

Then the semantics of counterfactual *could*, designated as $could^{CF}$, is:

(33)

$$S[\![\phi \; could^{CF} \; \psi]\!]^c =$$
$$\{\langle w, f\rangle \in S \mid$$
$$\text{If there is } w' \text{ such that Revises}(w', w, \mathcal{W}([\![\phi]\!]^c)), \text{ then}$$
$$\{\langle w', f\rangle \mid \text{Revises}(w', w, \mathcal{W}([\![\phi]\!]^c))\}[\![\phi]\!]^c[\![\psi]\!]^c \neq \emptyset\}$$

We assume a bare counterfactual exploits some contextually salient proposition or set of propositions to play the role of $\mathcal{W}([\![\phi]\!]^c)$ in the above definition. Let

(34) $[\![$ a river be near the location$_l$ of this utterance$_u$ $]\!]^c$

be the meaning of the consequent in (31b) evaluated against a context c. If we attempt to employ (33) to analyze (31b), using (34) for $[\![\psi]\!]^c$ and using a contextually salient set of worlds W_0 in place of of $\mathcal{W}([\![\phi]\!]^c)$, we will have an undefined result unless the input state already contains discourse markers l for the location of this utterance, and u for the utterance.

This is because of the kinds of states that (33) requires us to evaluate $[\![\psi]\!]^c$ against. Let $R(W, w, f)$ be:

$$\{\langle w', f\rangle \mid \text{Revises}(w', w, W)\}$$

for a worldset W and for any world w and assignment f. Then, using (33) without an overt antecedent means evaluating

$$R(W_0, w, f) \; [\![\psi]\!]^c$$

for the pragmatically given W_0 and for each w and f in the input state. In a counterfactual with an overt antecedent, the analogous update for each w and f would on the other hand be:

$$R(\mathcal{W}([\![\phi]\!]^c), w, f)[\![\phi]\!]^c[\![\psi]\!]^c$$

and $[\![\phi]\!]^c$ may thus provide an antecedent for any discourse marker in ψ. In the case of l and u in (34), this would allow for a reading on which the location of the utterance can vary from world to world, a reading such as we saw in (30a) and (30b). But in a context which fails to provide such a ϕ, the only way to get such a reading is through accommodation.

Let us assume the input state S already contains discourse marker d for the demonstration accompanying the demonstrative. Then for the reading of (31b) we are interested in, we need a state accommodating an utterance location related to that demonstration for each assignment

f. We can do this via:

$\Delta(S) =$
$\quad$ Accom(Accom(S, demonstratum-of, u, d),
$\qquad$ location-of, l, u) =
$\qquad \{\langle w, g \rangle \mid$
$\qquad\qquad$ There is $\langle w, f \rangle \in S$ and $f <_{u,l} g$
$\qquad\qquad$ and $\langle g(l), g(u) \rangle \in w$(location-of)
$\qquad\qquad$ and $\langle g(u), g(d) \rangle \in w$(demonstratum-of)$\}$

Even if d and u are rigid designators across revision worlds, the location of u may vary from revision world to revision world. Then if W_0, R, Δ, and $[\![\psi]\!]^c$ are as above and $could^{BCF}$ designates the bare counterfactual *could*,

(35) $\quad$ S$[\![could^{BCF} \; \psi]\!]^c =$
$\qquad \{\langle w, f \rangle \in S \mid$ If there is w' such that Revises(w', w, W_0),
$\qquad\qquad$ then $\Delta(R(W_0, w, f))[\![\psi]\!]^c \neq \emptyset\}$

gives the desired reading for (31b).

We now turn to (31a) and (31c). On our account the indexical readings of these sentences are available without any accommodation. This is because the Utterance Condition and the Consistency Condition taken together guarantee that any information state and context that are arguments of the meaning of (31a) and (31c) provide a discourse marker for all the contextual roles; and the utterance location is one of the contextual roles, namely the one filled by **here**$_c$.

The next question to ask is why the accommodation available for (31b) is not available for (31c). On our account, the implicit argument of *nearby* must have a corresponding discourse marker assigned to a location in the discourse. Why is it not possible to satisfy that requirement as before, by relating the implicit argument to the location of the utterance? In other words, using l again for the location and u for the utterance, why not accommodate l by way of location-of(l, u)?

The answer lies in the nature of ordinary accommodation. Accommodating a definite requires relating it to an entity already given in the discourse. But there is no reason in the case of a discourse in which u has not been overtly introduced, to assume that the utterance u is given. And in uttering (31c), in, say, a context where the speaker is evaluating the aesthetic merits of the utterance location, the utterance itself has *not* been introduced; in this case there is no demonstrative forcing the introduction of a discourse marker for the utterance.

Consider on the other hand the following sentences uttered in a context in which the speaker (perhaps in wistfully addressing himself) is evaluating the aesthetic merits of the utterance location:

(36) $\quad$ a. There could have been a river near me.

b. There could have been a river nearby.

In this case both (36a) and (36b) have readings on which alternative locations of the speaker are considered. Note that this is quite different from considering alternative locations for the utterance, since the speaker need not be uttering (36a) in the alternative locations considered. In the case of (36a) a reading on which alternative locations of the speaker are considered is possible without accommodation. Every state supplies a discourse marker for the speaker and (36a) is about counterfactual locations of that speaker. In the case of (36b), accommodation intervenes. The location required by *nearby* can be accommodated by way of the relation location-of(l,s) where s is the speaker. Accommodating by way of a relation to the speaker is possible in any context, because speaker is one of the contextual roles of an utterance context and we require every information state to supply a discourse marker for the speaker. In contrast, the accommodation in (32) was to the utterance u, and only a prior accommodation could guarantee the required discourse marker.

The fact that, in relevant contexts, (36b) can share the non-referential reading of (36a) but not the non-referential reading of (31b) shows that something distinguishes speakers from other necessary participants of the utterance context, such as the utterance itself. It thus provides important evidence for the idea of contextual roles. Any account that automatically accommodates whatever is always in the utterance context will accommodate too much and allow the impossible reading of (36b). Any account that omits speakers will accommodate too little.

4.4 Contextuals

We have argued that in dealing with the control of implicit arguments by the context of utterance we need to conceptually separate two properties, *indicativeness* and *referentiality*.

The canonical examples of indicative expressions are pronouns like *I* and demonstratives in their true indexical uses. The reference of an utterance of *I* is determined by whoever produces the utterance, but the relation of speaker and utterance is not part of the content of the sentence *I am hungry*. The reference of a true indexical use of *that* is the object pointed to, but the relation between pointer and pointed-at is not part of the content of an utterance.

Following Kaplan, we have chosen to pursue a direct reference account of the indicativeness of indexicals. This directly referential account carries over to our treatment of implicit arguments that anchor to the context of utterance.

Nunberg (1992) calls words like *local* contextuals. He defines contextuals as expressions that permit control by the context of utterance but whose meaning remains part of the utterance content, and points out that in cases like (37) they are controlled by the context of utterance without being referential.

(37)   a. The local scenery is getting prettier.
       b. The scenery around here is getting prettier.

Note that (37a) has a reading (37b) lacks: the discussion may be taking place on a train with the scenery moving. Here *local scenery* seems to vary with the motion of the train, but the interpretation is still controlled by the context of utterance. The fact that (37b) lacks a reading which allows *here* to vary is exactly what the direct reference account would predict, but is the quantification over locations in (37a) a problem for our analysis of implicit arguments?

We believe not. Our account of *local* allows the presuppositional requirements of the location argument to be satisified by accommodating a location into the input information state. As in the case of (36b), that accommodated location, however, may be controlled by the context of utterance without referring directly to the location of the utterance. Suppose for all w and f the location x is accommodated by the following relation:

$$\langle f(x), \mathbf{ego}_c \rangle \in w(\text{location-of})$$

Our account now requires that $\mathbf{ego}_c$ remain fixed, but not that the value of x remain fixed. In effect, we give this sentence the same account we would give:

(38)   The scenery around me is getting prettier.

Note that this sentence too has the reading which (37a) has and (37b) lacks.

We might also note in passing that according to Nunberg's definition of contextuals, a word like *we* is also a contextual. The relevant example is Partee's:

(39)   Every time a musician comes over, we play duets.

Here the reference of *we* may vary even though *we* remains anchored to a single utterance situation (a single speaker). If g is the group-denoting variable introduced by an utterance of *we*, then g will be restricted by the condition:

$$\langle \mathbf{ego}_c, f(g) \rangle \in w(\text{member-of})$$

To sum up, an implicit argument may be controlled by the context of utterance without being referential, as in (37a), but that case involves accommodation of a location-denoting discourse marker via the $\mathbf{ego}_c$ contextual role. In other words, that case is no different from a description containing an indexical and does not constitute a special kind of indexicality.

5 Conclusion

In this paper we have provided an account of the context-dependency of implicit arguments that results in a unified treatment of their readings and an account of indexicals that captures their indicativeness and referentiality. The two accounts are independent and their interaction falls out naturally. This is a desirable property for a theory relating anaphoric elements and indexicality to have. A wide range of linguistic elements can either function anaphorically or exploit the speech situation (third person pronouns, definites, implicit arguments, demonstratives). Thus we want anaphoric dependencies to be naturally satisfiable by features of the utterance situation. The notion of the context of utterance seems to be naturally extendable, by extending the domain of the context assignment function.

For the phenomena discussed in this paper, it would have been sufficient to assume that each information state has three designated variables whose unique values are **ego**, **here**, and **now**, respectively. Instead, we have introduced a context, a special kind of information state with maximal information with respect to worlds and the variables in its domain, for two main reasons. First, utterance situations may contain additional elements, which determine the interpretation of linguistic expressions such as demonstratives. Therefore, we want a construct that can in principle provide information about any number of relevant contextual features. The second reason has to do with the logic of indexicals: in order to capture the logical properties of certain sentences containing indexicals, such as pragmatic validity, we need to relativize interpretation to a context and an information state. These are, of course, among the principal reasons that motivated Kaplan's separation of the context of utterance from other parameters of evaluation and they apply to dynamic systems of semantic interpetation as much as they applied to classical truth-conditional systems of semantic interpetation.

References

Almog, Joseph, John Perry and Howard K. Wettstein, eds. 1989. *Themes from Kaplan*. New York: Oxford University Press.

Barwise, Jon and John Perry. 1983. *Situations and Attitudes*. Cambridge: MIT Press.

Beaver, David, I. 1993. *What Comes First in Dynamic Semantics: A Compositional, Non-Representational Account of Natural Language Presupposition*. Institute for Logic, Language and Computation. ILLC Prepublication Series for Logic, Semantics and Philosophy of Language LP-93-15. University of Amsterdam.

Chierchia, Gennaro. 1992. Anaphora and Dynamic Binding. *Linguistics and Philosophy* 15, 111–183.

Cresswell, Maxwell J. 1973. *Logics and Languages*. London: Methuen.

Crimmins, Mark D. 1992. *Talk About Beliefs*. Cambridge: A Bradford Book, The MIT Press.

Dekker, Paul J. E. 1993. *Transsentential Meditations: Ups and Downs in Dynamic Semantics*. Doctoral dissertation, University of Amsterdam. ILLC Dissertation Series, no. 1.

Dowty, David R. 1981. Quantification and the Lexicon: A Reply to Fodor and Fodor. In Michael Moortgat, Harry van der Hulst and Teun Hoekstra, eds., *The Scope of Lexical Rules*, 79–106. Dordrecht: Foris Publications.

Enç, Mürvet. 1986. Toward a Referential Analysis of Temporal Expressions. *Linguistics and Philosophy* 9, 405–426.

Fillmore, Charles J. 1969. Types of Lexical Information. In Ferenc Kiefer, ed., *Studies in Syntax and Semantics*, 109–137. Dordrecht: Reidel.

Fillmore, Charles J. 1986. Pragmatically Controlled Zero Anaphora. *Proceedings of the 14th Annual Meeting of the Berkeley Linguistics Society*, 35–55.

Fodor, Jerry A. and Janet D. Fodor. 1980. Functional Structure, Quantifiers, and Meaning Postulates. *Linguistic Inquiry* 11, 759–770.

Groenendijk, Jeroen A.G. & Martin J.B. Stokhof. 1991. Two Theories of Dynamic Semantics. In Jan van Eijck, ed., *Logics in AI: Proceedings of the European Workshop JELIA*, 55–64. (Lecture Notes in Computer Science, 478. Lecture Notes in Artificial Intelligence.) Berlin: Springer-Verlag.

Heim, Irene R. 1982. *The Semantics of Definite and Indefinite Noun Phrases*. Doctoral dissertation, University of Massachusetts, Amherst.

Heim, Irene R. 1983. On the Projection Problem for Presuppositions. In Michael Barlow, Daniel P. Flickinger, and Michael T. Wescoat, eds., *Proceedings of the West Coast Conference in Formal Linguistics*, vol. 2, 114–125. Stanford Linguistics Association.

Heim, Irene R. 1990. E-Type Pronouns and Donkey Anaphora. *Linguistics and Philosophy* 13, 137–177.

Hinrichs, Erhard. 1986. Temporal Anaphora in Discourses in English. *Linguistics and Philosophy* 9, 63–82.

Kamp, Hans. 1975. Two Theories About Adjectives. In Edward Keenan, ed., *Semantics for Natural Language*. Cambridge: Cambridge University Press.

Kamp, Hans. 1981. A Theory of Truth and Semantic Representation. In Jeroen Groenendijk, Theo Janssen, and Martin Stokhof, eds., *Formal Methods in the Study of Language*, 277–321. Amsterdam: Mathematisch Centrum.

Kamp, Hans. 1990. Prolegomena to a Structural Account of Belief and Other Attitudes. In C. Anthony Anderson and Joseph Owens, eds., *Proposi-*

tional Attitudes: The Role of Content in Logic, Language, and Mind, 27–90. CSLI Lecture Notes, Number 20. Stanford: CSLI Publications.

Kamp, Hans and Christian Rohrer. 1983. Tense in Texts. In Rainer Bäuerle, Christoph Schwarze and Arnim von Stechow, eds., *Meaning, Use and Interpretation of Language,* 250–269. Berlin: W. de Gruyter.

Kaplan, David. 1979. On the Logic of Demonstratives. In Peter A. French, Theodore E. Uehling Jr. and Howard K. Wettstein, eds., *Contemporary Perspectives on the Philosophy of Language,* 401–412. Minneapolis: University of Minnesota Press.

Kaplan, David. 1989a. Demonstratives: An Essay on the Semantics, Logic, Metaphysics, and Epistemology of Demonstratives and Other Indexicals. In Joseph Almog, John Perry and Howard K. Wettstein, eds., 481–563.

Kaplan, David. 1989b. Afterthoughts. In Joseph Almog, John Perry and Howard K. Wettstein, eds., 565–614.

Karttunen, Lauri. 1969. What Makes Definite Noun Phrases Definite? RAND Corporation Report No. P3871. Santa Monica, California.

Lewis, David K. 1973. *Counterfactuals.* Cambridge: Harvard University Press.

Lewis, David K. 1979. Score-keeping in a Language Game. In Rainer Bäuerle, Urs Egli and Arnim von Stechow, eds., *Semantics from Different Points of View,* 172–185. Berlin: Springer.

Lewis, David K. 1981. Index, Context, and Content. In Stig Kanger and Sven Öhman, eds., *Philosophy and Grammar,* 79–100. Synthese library, vol. 143. Dordrecht: Reidel.

Mitchell, Jonathan E. 1986. *The Formal Semantics of Point of View.* Doctoral dissertation, University of Massachusetts, Amherst.

Nunberg, Geoffrey. 1992. Two Kinds of Indexicality. In Chris Barker and David Dowty, eds., *Proceedings of the Second Conference on Semantics and Linguistic Theory,* 283–301. Columbus: Ohio State University.

Nunberg, Geoffrey. 1993. Indexicality and Deixis. *Linguistics and Philosophy* 16, 1–43.

Partee, Barbara H. 1984. Nominal and Temporal Anaphora. *Linguistics and Philosophy* 7, 243–286.

Partee, Barbara H. 1989. Binding Implicit Variables in Quantified Contexts. *Papers of the Chicago Linguistic Society,* volume 25, 342–365.

Perry, John. 1977. Frege on Demonstratives. *Philosophical Review* 86, 474–497.

Perry, John. 1986. Thoughts Without Representation. *Supplementary Proceedings of the Aristotelian Society* 60, 137–152.

Roberts, Craige. 1989. Modal Subordination and Pronominal Anaphora in Discourse. *Linguistics and Philosophy* 12:683–721.

Shopen, Timothy. 1973. Ellipsis as Grammatical Indeterminacy. *Foundations of Language* 10, 65–77.

Stalnaker, Robert C. 1968. A Theory of Conditionals. In Nicholas Rescher, ed., *Studies in Logical Theory,* 98–112. Oxford: Blackwell.

Stalnaker, Robert C. 1972. Pragmatics. In Donald Davidson and Gilbert Harman, eds., *Semantics of Natural Language*, 380–397. Dordrecht: Reidel.

Stalnaker, Robert C. 1978. Assertion. *Syntax and Semantics, Vol. 9: Pragmatics*, ed. Peter Cole, 315–332. New York: Academic Press.

Thomas, Andrew L. 1979. Ellipsis: The Interplay of Sentence Structure and Context. *Lingua* 47, 43-68.

2

A Deductive Account of Quantification in LFG

MARY DALRYMPLE, JOHN LAMPING, FERNANDO PEREIRA AND VIJAY SARASWAT

The relationship between Lexical-Functional Grammar (LFG) *functional structures* (f-structures) for sentences and their semantic interpretations can be expressed directly in a fragment of linear logic in a way that explains correctly the constrained interactions between quantifier scope ambiguity and bound anaphora.

The use of a deductive framework to account for the compositional properties of quantifying expressions in natural language obviates the need for additional mechanisms, such as Cooper storage, to represent the different scopes that a quantifier might take. Instead, the semantic contribution of a quantifier is recorded as an ordinary logical formula, one whose use in a proof will establish the scope of the quantifier. The properties of linear logic ensure that each quantifier is scoped exactly once.

Our analysis of quantifier scope can be seen as a recasting of Pereira's analysis (Pereira 1991), which was expressed in higher-order intuitionistic logic. But our use of LFG and linear logic provides a much more direct and computationally more flexible interpretation mechanism for at least the same range of phenomena. We have developed a

We thank Johan van Benthem, Bob Carpenter, Jan van Eijck, Angie Hinrichs, David Israel, Ron Kaplan, John Maxwell, Michael Moortgat, John Nerbonne, Stanley Peters, Henriette de Swart and an anonymous reviewer for discussions and comments. They are not responsible for any remaining errors, and we doubt that they will endorse all our analyses and conclusions, but we are sure that the end result is much improved for their help.

Quantifiers, Deduction, and Context
Makoto Kanazawa, Christopher Piñón, and Henriëtte de Swart, editors

preliminary Prolog implementation of the linear deductions described
in this work.

1 Introduction

This paper describes part of our ongoing investigation on the use of
formal deduction in linear logic to explicate the relationship between
syntactic analyses in Lexical-Functional Grammar (LFG) and semantic
interpretations. The use of formal deduction in semantic interpretation was implicit in deductive systems for categorial syntax (Lambek
1958), and has been made explicit through applications of the Curry-
Howard parallelism between proofs and terms in more recent work
on categorial semantics (van Benthem 1988, 1991), labeled deductive
systems (Moortgat 1992b) and flexible categorial systems (Hendriks
1993). Accounts of the syntax-semantics interface in the categorial tra-
dition require that syntactic and semantic analyses be formalized in
parallel algebraic structures of similar signatures, based on generalized
application and abstraction (or residuation) operators, and structure-
preserving relations between them. Those accounts therefore force the
adoption of categorial syntactic analyses, with their strong dependence
on phrase structure and linear order.

In contrast, our approach uses linear logic (Girard 1987) to represent
the connection between two dissimilar levels of representation, LFG f-
structures and their semantic interpretations. F-structures provide a
crosslinguistically uniform representation of syntactic information relevant to semantic interpretation that abstracts away from the details
of phrase structure and linear order in particular languages. This generality is in part achieved by using grammatical functions rather than
functor-argument relations to represent syntactic predicate-argument
relationships. As Halvorsen (1988) notes, however, the flatter, un-
ordered, grammatical function structure of LFG does not fit well with
traditional semantic compositionality, based on functional abstraction
and application, which mandates a rigid order of semantic composition.
We are thus forced to use a more relaxed form of compositionality, in
which, as in more traditional ones, the semantics of each lexical entry in a sentence is used exactly once in interpretation, but without
imposing a rigid order of composition. It turns out that linear logic offers exactly what is required for a calculus of semantic composition for
LFG, in that it can represent directly the constraints on the creation
and use of semantic units in sentence interpretation without forcing a
particular hierarchical order of composition except as required by the
properties of particular lexical entries.

We have shown previously that the linear-logic formalization of
the syntax-semantics interface for LFG provides simple and general

analyses of modification, functional completeness and coherence, and complex predicate formation (Dalrymple, Lamping and Saraswat 1993, Dalrymple et al. 1993). In the present paper, the analysis is extended to the interpretation of quantified noun phrases. After an overview of the approach, we present our analysis of the compositional properties of quantifiers, and we conclude by showing that the analysis correctly accounts for scope ambiguity and its interactions with bound anaphora.

2 LFG and Linear Logic

Syntactic Framework LFG assumes two syntactic levels of representation: constituent structure (*c-structure*) represents phrasal dominance and precedence relations, while functional structure (*f-structure*) represents syntactic predicate-argument structure. For example, the f-structure for sentence (1) is given in (2):

(1)   Bill appointed Hillary.

(2) $$\begin{bmatrix} \text{PRED} & \text{`appoint'} \\ \text{SUBJ} & \begin{bmatrix} \text{PRED} & \text{`Bill'} \end{bmatrix} \\ \text{OBJ} & \begin{bmatrix} \text{PRED} & \text{`Hillary'} \end{bmatrix} \end{bmatrix}$$

As illustrated, a functional structure consists of a collection of attributes, such as PRED, SUBJ, and OBJ, whose values can, in turn, be other functional structures. The following annotated phrase-structure rules can generate the f-structure in (2):

(3)   S $\longrightarrow$ NP   VP
                 ($\uparrow$ SUBJ) = $\downarrow$ $\uparrow$= $\downarrow$
        VP $\longrightarrow$ V   NP
                 $\uparrow$= $\downarrow$ ($\uparrow$ OBJ) = $\downarrow$

These two phrase structure rules do not encode semantic information; they specify only how grammatical functions such as SUBJ are expressed in English. The f-structure metavariables $\uparrow$ and $\downarrow$ refer, respectively, to the f-structure of the mother of the current node and to the f-structure of the current node (Kaplan and Bresnan 1982). The annotations on the S rule indicate, then, that the f-structure for the S has a SUBJ attribute whose value is the f-structure for the NP daughter, and that the f-structure for the S is the same as the one for the VP daughter. The relation between the nodes of the c-structure and the f-structure for the sentence (1) is expressed by means of arrows in (4):

(4)

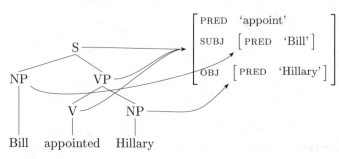

Lexically-Specified Semantics Unlike phrase structure rules, lexical entries specify semantic as well as syntactic information. Here are the lexical entries for the words in the sentence:

(5)   *Bill*   NP   ($\uparrow$ PRED) = 'Bill'
                   $\uparrow_\sigma \leadsto Bill$
     appointed   V   ($\uparrow$ PRED)= 'appoint'
       $\forall X, Y. (\uparrow \text{SUBJ})_\sigma \leadsto X \otimes (\uparrow \text{OBJ})_\sigma \leadsto Y \multimap \uparrow_\sigma \leadsto appoint(X, Y)$
     Hillary   NP   ($\uparrow$ PRED) = 'Hillary'
                   $\uparrow_\sigma \leadsto Hillary$

Just like phrase structure rules, lexical entries are instantiated for a particular utterance. The metavariable $\uparrow$ in a lexical entry represents the f-structure of the c-structure mother of (an instance of) the entry in a c-structure. The syntactic information given in lexical entries consists of equality statements about the f-structure, while the semantic information consists of assertions about how the meaning of the f-structure participates in various semantic relations.

The semantic information in a lexical entry, which we will call the *semantic contribution* of the entry, is a linear-logic formula that constrains the association between *semantic structures* projected from the f-structures mentioned in the lexical entry (Kaplan 1987, Halvorsen and Kaplan 1988) and their semantic interpretations. The semantic projection function σ maps an f-structure to a semantic structure encoding information about its meaning, in the same way as the functional projection function ϕ maps c-structure nodes to the associated f-structures.

The association between f_σ and a meaning P is represented by the atomic formula $f_\sigma \leadsto P$, where $\leadsto$ is an otherwise uninterpreted binary predicate symbol. (In fact, we use not one but a family of relations $\leadsto_\tau$ indexed by the semantic type of the intended second argument, although for simplicity we will omit the type subscript whenever it is determinable from context.) We will often informally say that P is f's meaning without referring to the role of the semantic structure f_σ in $f_\sigma \leadsto P$. We will see, however, that f-structures and their semantic projections must be distinguished, because in general semantic projections carry more information than just the association to the meaning for the corresponding f-structure.

We can now explain the semantic contributions in (5). If a particular occurrence of 'Bill' in a sentence is associated with f-structure f, the syntactic constraint in the lexical entry *Bill* will be instantiated as $(f \text{ PRED}) = $ 'Bill' and the semantic constraint will be instantiated as $f_\sigma \rightsquigarrow Bill$, representing the association between f_σ and the constant *Bill* representing its meaning.

The semantic contribution of the *appointed* entry is more complex, as it relates the meanings of the subject and object of a clause to the clause's meaning. Specifically, if f is the f-structure for a clause with predicate (PRED) 'appoint', the semantic contribution asserts that if f's subject $(f \text{ SUBJ})$ has meaning X and (*multiplicative* conjunction $\otimes$) f's object $(f \text{ OBJ})$ has meaning Y, then (linear implication $\multimap$) f has meaning $appoint(X, Y)$.[1]

Logical Representation of Semantic Compositionality In the semantic contribution for *appointed* in (5), the linear-logic connectives of multiplicative conjunction $\otimes$ and linear implication $\multimap$ are used to specify how the meaning of a clause headed by the verb is composed from the meanings of the arguments of the verb. For the moment, we can think of the linear connectives as playing the same role as the analogous classical connectives of conjunction and implication, but we will soon see that the specific properties of the linear connectives are essential to guarantee that lexical entries bring into the interpretation process all and only the information provided by the corresponding words. The semantic contribution of *appointed* asserts that if the subject of a clause with main verb *appointed* means X and its object means Y, then the whole clause means $appoint(X, Y)$. The semantic contribution can thus be thought of as a linear definite clause, with the variables X and Y playing the same role as Prolog variables.

---

[1] In fact, we believe that the correct treatment of the relation between a verb and its arguments requires the use of *mapping principles* specifying the relation between the array of semantic arguments required by a verb and their possible syntactic realizations (Bresnan and Kanerva 1989, Alsina 1993, Butt 1993). A verb like *appoint*, for example, might specify that one of its arguments is an agent and the other is a theme. Mapping principles would then specify that agents can be realized as subjects and themes as objects.

Here we make the simplifying assumption that the arguments of verbs have already been linked to syntactic functions and that this linking is represented in the lexicon, since for the examples we will discuss this assumption is innocuous. However, in the case of *complex predicates* this assumption produces incorrect results, as shown by Butt (1993). Mapping principles are very naturally incorporated into the framework discussed here; see Dalrymple, Lamping, and Saraswat (1993) and Dalrymple et al. (1993) for discussion and illustration.

It is worth noting that the form of the semantic contribution of [2] *appointed* parallels the type $e \times e \to t$ which, in its curried form $e \to e \to t$, is the standard type for a transitive verb in a compositional semantics setting (Gamut 1991). In general, the propositional structure of the semantic contributions of lexical entries will parallel the types assigned to the meanings of the same words in compositional analyses.

Given the semantic contributions in (5), we can derive deductively the meaning for example (1). Let the constants f, g and h name the following f-structures:

(6)
$$f: \begin{bmatrix} \text{PRED} & \text{'appoint'} \\ \text{SUBJ} & g: [\text{PRED} \quad \text{'Bill'}] \\ \text{OBJ} & h: [\text{PRED} \quad \text{'Hillary'}] \end{bmatrix}$$

Instantiating the lexical entries for *Bill*, *Hillary*, and *appointed* appropriately, we obtain the following semantic contributions, abbreviated as **bill**, **hillary**, and **appointed**:

bill: $g_\sigma \rightsquigarrow Bill$
hillary: $h_\sigma \rightsquigarrow Hillary$
appointed: $\forall X, Y. \ g_\sigma \rightsquigarrow X \otimes h_\sigma \rightsquigarrow Y \multimap f_\sigma \rightsquigarrow appoint(X, Y)$

These formulas show how the generic semantic contributions in the lexical entries are instantiated to reflect their participation in this particular f-structure. For example, since the entry *Bill* is used for f-structure g, the semantic contribution for *Bill* provides a meaning for g_σ. More interestingly, the verb *appointed* requires two pieces of information, the meanings of its subject and object, in no particular order, to produce a meaning for the clause. As instantiated, the f-structures corresponding to the subject and object of the verb are g and h, respectively, and f is the f-structure for the entire clause. Thus, the instantiated entry for *appointed* shows how to combine a meaning for g_σ (its subject) and h_σ (its object) to generate a meaning for f_σ (the entire clause).

In the following, assume that the formula **bill-appointed** is defined thus:

bill-appointed: $\forall Y. \ h_\sigma \rightsquigarrow Y \multimap f_\sigma \rightsquigarrow appoint(Bill, Y)$

Then the following derivation is possible in linear logic ($\vdash$ stands for the linear-logic entailment relation):

(7)   **bill** $\otimes$ **hillary** $\otimes$ **appointed**   (*Premises.*)
   $\vdash$ **bill-appointed** $\otimes$ **hillary**
   $\vdash$ $f_\sigma \rightsquigarrow appoint(Bill, Hillary)$

At each step, universal instantiation and modus ponens are used.

In summary, each word in a sentence contributes a linear-logic formula relating the semantic projections of specific f-structures in the

LFG analysis to representations of their meanings. From those formulas, the interpretation process attempts to deduce an atomic formula relating the semantic projection of the whole sentence to a representation of the sentence's meaning. Alternative derivations may yield different such conclusions, corresponding to semantic interpretation ambiguities.

Meaning and glue Our approach shares the order-independence of representations of semantic information by attribute-value matrices (Pollard and Sag 1987, Fenstad et al. 1987, Pollard and Sag 1993), while still allowing a well-defined treatment of variable binding and scope. We do this by distinguishing (1) a *language of meanings* and (2) a language for assembling meanings or *glue language*.

The language of meanings could be that of any appropriate logic, for instance Montague's intensional logic (Montague 1974). The glue language, described below, is a fragment of linear logic. The semantic contribution of each lexical entry is represented by a linear-logic formula that can be understood as instructions in the glue language for combining the meanings of the lexical entry's syntactic arguments into the meaning of the f-structure headed by the entry. Glue formulas may also be contributed by some syntactic constructions, when properties of a construction as a whole and not just of its lexical elements are responsible for the interpretation of the construction; these cases include the semantics of relative clauses. We will not discuss construction-specific interpretation rules in this paper.

Appendix 4 gives further details on the syntax of the meaning and glue languages used in this paper.

Linear logic As we have just outlined, we use deduction in linear logic to assign meanings to sentences, starting from information about their functional structure and about the semantics of the words they contain. An approach based on linear logic, which crucially allows premises to commute, appears to be more compatible with the shallow and relatively free-form functional structure than are compositional approaches, which rely on deeply nested binary-branching immediate dominance relationships. As noted above, the use of linear logic as the system for assembling meanings permits a uniform treatment of a range of natural language phenomena described by Dalrymple, Lamping, and Saraswat (1993), including modification, completeness and coherence,[2] and complex predicate formation.

---

[2] "An f-structure is *locally complete* if and only if it contains all the governable grammatical functions that its predicate governs. An f-structure is *complete* if and only if all its subsidiary f-structures are locally complete. An f-structure is *locally coherent* if and only if all the governable grammatical functions that it contains are governed by a local predicate. An f-structure is *coherent* if and only if all its subsidiary f-structures are locally coherent." (Kaplan and Bresnan 1982, pages 211–212).

An important motivation for using linear logic is that it allows us to capture directly the intuition that lexical items and phrases each contribute exactly once to the meaning of a sentence. As noted by Klein and Sag (1985, page 172):

> Translation rules in Montague semantics have the property that the translation of each component of a complex expression occurs exactly once in the translation of the whole. ...That is to say, we do not want the set S [of semantic representations of a phrase] to contain *all* meaningful expressions of IL which can be built up from the elements of S, but only those which use each element exactly once.

In our terms, the semantic contributions of the constituents of a sentence are not context-independent assertions that may be used or not in the derivation of the meaning of the sentence depending on the course of the derivation. Instead, the semantic contributions are *occurrences* of information which are generated and used exactly once. For example, the formula $g_\sigma \leadsto Bill$ can be thought of as providing one occurrence of the meaning *Bill* associated to the semantic projection g_σ. That meaning must be consumed exactly once (for example, by **appointed** in (7)) in the derivation of a meaning of the entire utterance.

It is this "resource-sensitivity" of natural language semantics—an expression is used exactly once in a semantic derivation—that linear logic can model. The basic insight underlying linear logic is that logical formulas are *resources* that are produced and consumed in the deduction process. This gives rise to a resource-sensitive notion of implication, the *linear implication* $\multimap$: the formula $A \multimap B$ can be thought of as an action that can *consume* (one copy of) A to produce (one copy of) B. Thus, the formula $A \otimes (A \multimap B)$ linearly entails B. It does not entail $A \otimes B$ (because the deduction consumes A), and it does not entail $(A \multimap B) \otimes B$ (because the linear implication is also consumed in doing the deduction). This resource-sensitivity not only disallows arbitrary duplication of formulas, but also disallows arbitrary deletion of formulas. Thus multiplicative conjunction $\otimes$ is sensitive to the multiplicity of formulas: $A \otimes A$ is not equivalent to A (the former has two copies of the formula A). For example, the formula $A \otimes A \otimes (A \multimap B)$ linearly entails $A \otimes B$ (there is still one A left over) but does not entail B (there must still be one A present). In this way, linear logic checks that a formula is used once and only once in a deduction, enforcing the requirement that each component of an utterance contributes exactly once to the assembly of the utterance's meaning.

To handle quantification, our glue language needs to be only a fragment of higher-order linear logic, the *tensor fragment*, that is closed under conjunction, universal quantification, and implication (with at most one level of nesting of implication in antecedents). In fact, all but the determiner lexical entries are in the first-order subset of this fragment. This fragment arises from transferring to linear logic the ideas underlying the concurrent constraint programming scheme of Saraswat (1989). An explicit formulation for the higher-order version of the linear concurrent constraint programming scheme is given in Saraswat and Lincoln (1992). A nice tutorial introduction to linear logic itself may be found in Scedrov (1993); see also Saraswat (1993).

Relationship with Categorial Syntax and Semantics As suggested above, there are interesting connections between our approach and various systems of categorial syntax and semantics. The Lambek calculus (Lambek 1958), introduced as a logic of syntactic combination, turns out to be a fragment of noncommutative multiplicative linear logic. If permutation is added to Lambek's system, its left- and right-implication connectives (\ and /) collapse into a single implication connective with behavior identical to $-\circ$. This undirected version of the Lambek calculus was developed by van Benthem (1988, 1991) to account for the semantic combination possibilities of phrase meanings.

Those systems and related ones (Moortgat 1988, Hepple 1990, Morrill 1990) were developed as calculi of syntactic/semantic types, with propositional formulas representing syntactic categories or semantic types. Given the types for the lexical items in a sentence as assumptions, the sentence is syntactically well-formed in the Lambek calculus if the type of the sentence can be derived from the assumptions arranged as an ordered list. Furthermore, the Curry-Howard isomorphism between proofs and terms (Howard 1980) allows the extraction of a term representing the meaning of the sentence from the proof that the sentence is well-formed (van Benthem 1986). However, the Lambek calculus and its variants carry with them a particular view of syntactic structure that is not obviously compatible with the flatter f-structures proposed by LFG. ˊ

On the other hand, categorial semantics in the undirected Lambek calculus and other related commutative calculi provides an analysis of the possibilities of meaning combination independently of the syntactic realizations of those meanings, but does not provide a mechanism for relating semantic combination possibilities to the corresponding syntactic combination possibilities.

In more recent work, multidimensional and labeled deductive systems (Moortgat 1992b, Morrill 1994) have been proposed as refinements of the Lambek systems that are able to represent synchronized derivations involving multiple levels of representation, for instance a level of head-dependent representations and a level of syntactic functor-argument representations. However, these systems do not yet seem able to represent the connection between a flat syntactic representation in terms of grammatical functions and a function-argument semantic representation. As far as we can see, the problem in those systems is that at the type level it is not possible to express the link between particular syntactic structures (f-structures in our case) and particular contributions to meaning. The extraction of meanings from derivations following the Curry-Howard isomorphism that is standard in categorial systems demands that the order of syntactic combination coincide with the order of semantic combination so that functor-argument relations at the syntactic and semantic level are properly aligned.

Thus, while the "propositional skeleton" of an analysis in our system can be seen as a close relative of the corresponding categorial semantics derivation in the undirected Lambek calculus, the first-order part of our analysis (notably the f, g, and h in the example above) explicitly carries the connection between f-structures and their contributions to meaning. In this way, we can take advantage of the principled description of potential meaning combinations of categorial semantics without losing track of the constraints imposed by syntax on the possible combinations of those meanings.

3 Quantification

Our treatment of quantification, and in particular of quantifier scope ambiguity and of the interactions between scope and bound anaphora, follows the approach of Pereira (1990, 1991). It turns out, however, that the linear-logic formulation is simpler and easier to justify than the earlier analysis, which used an intuitionistic type assignment logic.

The basic idea for the analysis can be seen as a logical counterpart at the glue level of the standard type assignment for generalized quantifiers (Barwise and Cooper 1981). The generalized quantifier meaning of a natural language determiner has the following type, a function from two properties, the quantifier's restriction and scope, to a proposition:

(8) $(e \to t) \to (e \to t) \to t$

At the semantic glue level, we can understand that type as follows. For any determiner, if for arbitrary x we can construct a meaning Rx for the quantifier's restriction, and again for arbitrary x we can construct a meaning Sx for the quantifier's scope, where R and S are properties (functions from entities to propositions), then we can construct the meaning QRS for the whole sentence containing the determiner, where Q is the meaning of the determiner. In the following we will notate QRS meaning more perspicuously as $Q(z, Rz, Sz)$.

Assume that we have determined the following semantic structures: *restr* for the restriction (a common noun phrase), *restr-arg* for its implicit argument, *scope* for the scope of quantification, and *scope-arg* for the grammatical function filled by the quantified noun phrase. Then the foregoing analysis can be represented in linear logic by the following schematic formula:

(9) $\forall R, S.\ (\forall x.\ restr\text{-}arg \leadsto x \multimap restr \leadsto Rx)$
     $\otimes (\forall x.\ scope\text{-}arg \leadsto x \multimap scope \leadsto Sx)$
     $\multimap scope \leadsto Q(z, Rz, Sz)$

Given the equivalence between $A \otimes B \multimap C$ and $A \multimap (B \multimap C)$, the propositional part of (9) parallels the generalized quantifier type (8).

In addition to providing a semantic type assignment for determiners, (9) uses glue language quantification to express how the meanings of the restriction and scope of quantification are determined and combined into the meaning of the quantified clause. The condition $\forall x.\ restr\text{-}arg \leadsto x \multimap restr \leadsto Rx$ specifies that, if for arbitrary x $restr\text{-}arg$ has meaning x,[3] then $restr$ has meaning Rx, that is, it gives the dependency of the meaning of a common noun phrase on its implicit argument. Property R is the representation of that dependency as a function. Similarly, the subformula $\forall x.\ scope\text{-}arg \leadsto x \multimap scope \leadsto Sx$ specifies the dependency of the meaning Sx of a semantic structure $scope$ on the meaning x of one of its arguments $scope\text{-}arg$. If both dependencies hold, then R and S are an appropriate restriction and scope for the determiner meaning Q.

Computationally, the nested universal quantifiers substitute unique new constants (eigenvariables) for the quantified variable x, and the nested implications try to prove their consequent with the antecedent added to the current set of assumptions. Higher-order unification (in this case, a fairly restricted version thereof (Miller 1990)) is used to solve for the values of R and S that satisfy the nested implication, which cannot contain occurrences of the eigenvariables.

To complete our specification of the semantic contribution of a determiner, we need to see how it relates to the f-structure it contributes to. The f-structure for a quantified noun phrase has the general form

(10) $f\!:\begin{bmatrix} \text{SPEC} & q \\ \text{PRED} & g\!:\cdots \end{bmatrix}$

---

[3]We use lower-case letters for *essentially universal* variables, that is, variables that stand for new local constants in a proof. We use capital letters for *essentially existential* variables, that is, Prolog-like variables that become instantiated to particular terms in a proof. In other words, essentially existential variables stand for specific but as yet unspecified terms, while essentially universal variables stand for arbitrary constants, that is, constants that could be replaced by *any* term while still maintaining the validity of the derivation. In the linear-logic fragment we use here, essentially existential variables arise from universal quantification with outermost scope, while essentially universal variables arise from universal quantification whose scope is a conjunct in the antecedent of an outermost implication.

where SPEC is the determiner and PRED is the noun.

Since the meaning of a noun is a property (type $e \to t$), its semantic contribution has the form of an implication, just like a verb.[4] However, while for a verb the arguments and result of the verb meaning can be associated to projections of appropriate f-structures in the syntactic analysis of the verb's clause, there are no appropriate f-structures in the analysis of a noun phrase that can be associated with the argument and result of the noun's meaning. Instead, we take the semantic projection f_σ of the noun phrase to be structured with two attributes (f_σ VAR) and (f_σ RESTR), performing a comparable role to the attributes CM and PM in the semantic structure in Halvorsen's (1983) treatment of quantifiers. Using those attributes, the semantic contribution of a noun can be expressed in the form

$$\forall x. (\uparrow_\sigma \text{ VAR}) \rightsquigarrow x \multimap (\uparrow_\sigma \text{ RESTR}) \rightsquigarrow Px$$

where P is the meaning of the noun.

We can now describe how the semantic structures *restr-arg*, *restr*, *scope-arg* and *scope* in (9) relate to the f-structure. The contribution of determiner q is expressed in terms of f's semantic projection f_σ. To connect the restriction of the determiner with the noun, we take *restr-arg* = (f_σ VAR) and *restr* = (f_σ RESTR). Since f fills an appropriate argument position, *scope-arg* = f_σ. As for *scope*, the scope of a determiner is not explicitly given, so we can only say that it can be any semantic structure,[5] subject to the constraint that the meaning associated to the semantic structure have proposition type t, that is, the semantic contribution should quantify universally over possible scopes. Therefore, the contribution of a determiner is:[6]

---

[4]For a discussion of relational nouns, whose meanings are relations rather than properties, see Section 3.4.

[5]Because of this indeterminacy as to choice of scope, Halvorsen and Kaplan (1988) used *inside-out functional uncertainty* to nondeterministically choose a scope f-structure for a quantifier. Their approach requires the scope of a quantifier to be an f-structure which contains the quantifier f-structure. In contrast, our approach places no syntactic constraints whatever on the choice of quantifier scope, since the propositional structure of the formulas involved in the derivation will preclude all but the appropriate choices.

[6]There is an alternative formulation of quantifier meaning that doesn't use nested implications:

$$\exists x. (\uparrow_\sigma \text{ VAR}) \rightsquigarrow x$$
$$\otimes \forall H, R. (\uparrow_\sigma \text{ RESTR}) \rightsquigarrow Rx \multimap (\uparrow_\sigma \rightsquigarrow x \otimes \forall S. (H \rightsquigarrow Sx) \multimap H \rightsquigarrow Q(z, Rz, Sz))$$

This formulation just asserts that there is a generic entity, x, which stands for the meaning of the quantified phrase, and also serves as the argument of the restriction. The derivations of the restriction and scope are then expected to consume this information. By avoiding nested implications, this formulation may be easier to work with computationally. However, the logical structure of this formulation is not as restrictive as that of (11), as it can allow additional derivations where information intended for the restriction can be used by the scope. In fact, though, for many

(11) $\quad \forall H, R, S.\ (\forall x.\ (\uparrow_\sigma \text{VAR}) \rightsquigarrow x \multimap (\uparrow_\sigma \text{RESTR}) \rightsquigarrow Rx)$
$$\otimes (\forall x.\ \uparrow_\sigma \rightsquigarrow x \multimap H \rightsquigarrow_t Sx)$$
$$\multimap H \rightsquigarrow_t Q(z, Rz, Sz)$$

where H ranges over semantic structures.

The VAR and RESTR components of the semantic projection for a quantified noun phrase in our analysis play a similar role to the $/\!/$ category constructor in PTQ (Montague 1974), that of distinguishing syntactic configurations with identical semantic types but different contributions to the interpretation. The two PTQ syntactic categories t/e for intransitive verb phrases and $t/\!/e$ for common noun phrases correspond to the single semantic type $e \rightarrow t$; similarly, the two conjuncts in the antecedent of (11) correspond to the same semantic type, encoded with a linear implication, but to two different syntactic contexts, one relating the predication of a noun phrase to its implicit argument and one relating a clause to an embedded argument.

3.1 Quantified noun phrase meanings

We first demonstrate how the semantic contribution of a quantified noun phrase such as *every voter* is derived. The following annotated phrase structure rule is necessary:

(12) $\quad$ NP $\longrightarrow$ $\quad$ Det $\qquad$ N
$$\qquad\qquad\qquad\quad \uparrow = \downarrow \qquad \uparrow = \downarrow$$

This rule states that the determiner Det and noun N each contribute to the f-structure for the NP. Lexical specifications ensure that the noun contributes the PRED attribute and its value, and the determiner contributes the SPEC attribute and its value. The f-structure for the noun phrase *every voter* is:

(13)
$$h:\begin{bmatrix} \text{SPEC} & \text{`every'} \\ \text{PRED} & \text{`voter'} \end{bmatrix}$$

The lexical entries used in this f-structure are:

(14) $\quad$ *every* $\quad$ Det $\quad$ ($\uparrow$ SPEC) = 'every'
$$\forall H, R, S.\ (\forall x.\ (\uparrow_\sigma \text{VAR}) \rightsquigarrow x \multimap (\uparrow_\sigma \text{RESTR}) \rightsquigarrow Rx)$$
$$\otimes (\forall x.\ \uparrow_\sigma \rightsquigarrow x \multimap H \rightsquigarrow Sx)$$
$$\multimap H \rightsquigarrow every(z, Rz, Sz)$$

(15) $\quad$ *voter* $\quad$ N $\quad$ ($\uparrow$ PRED) = 'voter'
$$\forall X.\ (\uparrow_\sigma \text{VAR}) \rightsquigarrow X \multimap (\uparrow_\sigma \text{RESTR}) \rightsquigarrow voter(X)$$

The semantic contributions of common nouns and determiners were described in the previous section.

---

analyses (including all those that we have investigated) such interactions are already precluded by the quantificational structure of the formula, and in such cases the formulation above is equivalent to (11).

Given those entries, the semantic contributions of *every* and *voter* in (13) are

every: $\forall H, R, S. \ (\forall x. \ (h_\sigma \ \text{VAR}) \leadsto x \ \multimap (h_\sigma \ \text{RESTR}) \leadsto Rx)$
$\otimes (\forall x. \ h_\sigma \leadsto x \ \multimap H \leadsto Sx)$
$\multimap \ H \leadsto every(z, Rz, Sz)$

voter: $\forall X. \ (h_\sigma \ \text{VAR}) \leadsto X \ \multimap (h_\sigma \ \text{RESTR}) \leadsto voter(X)$

From these two premises, the semantic contribution for *every voter* follows:

every-voter: $\forall H, S. \ (\forall x. \ h_\sigma \leadsto x \ \multimap H \leadsto Sx)$
$\multimap \ H \leadsto every(z, voter(z), Sz)$

The propositional part of this contribution corresponds to the standard type for noun phrase meanings, $(e \to t) \to t$. Informally, the whole contribution can be read as follows: if by giving the arbitrary meaning x of type e to the argument position filled by the noun phrase we can derive the meaning Sx of type t for the semantic structure scope of quantification H, then S can be the property that the noun phrase meaning requires as its scope, yielding the meaning $every(z, voter(z), Sz)$ for H. The quantified noun phrase can thus be seen as providing two contributions to an interpretation: locally, a *referential import* x, which must be discharged when the scope of quantification is established; and globally, a *quantificational import* of type $(e \to t) \to t$, which is applied to the meaning of the scope of quantification to obtain a quantified proposition.

3.2 Simple example of quantification

Before we look at quantifier scope ambiguity and interactions between scope and bound anaphora , we demonstrate the basic operation of our proposed representation of the semantic contribution of a determiner. We use the following sentence with a single quantifier and no scope ambiguities:

(16) Bill convinced every voter.

To carry out the analysis, we need a lexical entry for *convinced*:

(17)

     convinced V $(\uparrow \text{PRED}) = \text{`convince'}$
                    $(\uparrow \text{TENSE}) = \text{PAST}$
   $\forall X, Y. \ (\uparrow \text{SUBJ})_\sigma \leadsto X \otimes (\uparrow \text{OBJ})_\sigma \leadsto Y \ \multimap \uparrow_\sigma \leadsto convince(X, Y)$

The f-structure for (16) is:

(18)
$$f: \begin{bmatrix} \text{PRED} & \text{'convince'} \\ \text{TENSE} & \text{PAST} \\ \text{SUBJ} & g: \begin{bmatrix} \text{PRED} & \text{'Bill'} \end{bmatrix} \\ \text{OBJ} & h: \begin{bmatrix} \text{SPEC} & \text{'every'} \\ \text{PRED} & \text{'voter'} \end{bmatrix} \end{bmatrix}$$

The premises for the derivation are the semantic contributions for *Bill* and *convinced* together with the contribution derived above for the quantified noun phrase *every voter*:

bill: $\quad\quad\quad g_\sigma \leadsto Bill$

convinced: $\quad \forall X, Y. \; g_\sigma \leadsto X \otimes h_\sigma \leadsto Y \multimap f_\sigma \leadsto convince(X, Y)$

every-voter: $\quad \forall H, S. \; (\forall x. \; h_\sigma \leadsto x \multimap H \leadsto Sx)$
$$\multimap H \leadsto every(z, voter(z), Sz)$$

Giving the name **bill-convinced** to the formula

bill-convinced: $\quad \forall Y. \; h_\sigma \leadsto Y \multimap f_\sigma \leadsto convince(Bill, Y)$

we have the derivation:

$$\begin{array}{ll} & \textbf{bill} \otimes \textbf{convinced} \otimes \textbf{every-voter} \quad \text{(Premises.)} \\ \vdash & \textbf{bill-convinced} \otimes \textbf{every-voter} \\ \vdash & f_\sigma \leadsto every(z, voter(z), convince(Bill, z)) \end{array}$$

No derivation of a different formula $f_\sigma \leadsto_t P$ is possible. The formula **bill-convinced** represents the semantics of the scope of the determiner 'every'. The derivable formula

$$\forall Y. \; h_\sigma \leadsto_e Y \multimap h_\sigma \leadsto_e Y$$

could at first sight be considered another possible, but erroneous, scope. However, the type subscripting of the $\leadsto$ relation used in the determiner lexical entry requires the scope to represent a dependency of a proposition on an individual, while this formula represents the dependency of an individual on an individual (itself). Therefore, it does not provide a valid scope for the quantifier.

3.3 Quantifier scope ambiguities

When a sentence contains more than one quantifier, scope ambiguities are of course possible. In our system, those ambiguities will appear as alternative successful derivations. We will take as our example the sentence[7]

(19) Every candidate appointed a manager

for which we need the additional lexical entries

---

[7] In order to allow for apparent scope ambiguities, we adopt a scoping analysis of indefinites, as proposed, for example, by Neale (1990).

(20)

a Det $(\uparrow \text{SPEC}) = \text{'a'}$

$$\forall H, R, S. \ (\forall x. \ (\uparrow_\sigma \text{VAR}) \rightsquigarrow x \multimap (\uparrow_\sigma \text{RESTR}) \rightsquigarrow Rx)$$
$$\otimes (\forall x. \ \uparrow_\sigma \rightsquigarrow x \multimap H \rightsquigarrow Sx)$$
$$\multimap H \rightsquigarrow a(z, Rz, Sz)$$

(21)   *candidate*   N $(\uparrow \text{PRED}) = \text{'candidate'}$

$$\forall X. \ (\uparrow_\sigma \text{VAR}) \rightsquigarrow X \multimap (\uparrow_\sigma \text{RESTR}) \rightsquigarrow candidate(X)$$

(22)   *manager*   N $(\uparrow \text{PRED}) = \text{'manager'}$

$$\forall X. \ (\uparrow_\sigma \text{VAR}) \rightsquigarrow X \multimap (\uparrow_\sigma \text{RESTR}) \rightsquigarrow manager(X)$$

The f-structure for sentence (19) is

(23)

$$f: \begin{bmatrix} \text{PRED} & \text{'appoint'} \\ \text{TENSE} & \text{PAST} \\ \text{SUBJ} & g: \begin{bmatrix} \text{SPEC} & \text{'every'} \\ \text{PRED} & \text{'candidate'} \end{bmatrix} \\ \text{OBJ} & h: \begin{bmatrix} \text{SPEC} & \text{'a'} \\ \text{PRED} & \text{'manager'} \end{bmatrix} \end{bmatrix}$$

We can derive semantic contributions for *every candidate* and *a manager* in the way shown in Section 3.1. Further derivations proceed from those contributions together with the contribution of *appointed*:

every-candidate: $\forall H, S. \ (\forall x. \ g_\sigma \rightsquigarrow x \multimap H \rightsquigarrow Sx)$
$$\multimap H \rightsquigarrow every(w, candidate(w), Sw)$$

a-manager: $\forall H, S. \ (\forall x. \ h_\sigma \rightsquigarrow x \multimap H \rightsquigarrow Sx)$
$$\multimap H \rightsquigarrow a(z, manager(z), Sz)$$

appointed: $\forall X, Y. \ g_\sigma \rightsquigarrow X \otimes h_\sigma \rightsquigarrow Y \multimap f_\sigma \rightsquigarrow appoint(X, Y)$

As of yet, we have not made any commitment about the scopes of the quantifiers; the $\forall S$'s have not been instantiated. Scope ambiguities are manifested in two different ways in our system: through the choice of different semantic structures H, corresponding to different syntactic choices for where to scope the quantifier, or through different relative orders of quantifiers that scope at the same point. For this example, the second case is relevant, and we must now make a choice to proceed. The two possible choices correspond to two equivalent rewritings of **appointed**:

$$\forall X. \ g_\sigma \rightsquigarrow X \multimap (\forall Y. \ h_\sigma \rightsquigarrow Y \multimap f_\sigma \rightsquigarrow appoint(X, Y))$$
$$\forall Y. \ h_\sigma \rightsquigarrow Y \multimap (\forall X. \ g_\sigma \rightsquigarrow X \multimap f_\sigma \rightsquigarrow appoint(X, Y))$$

These two equivalent forms correspond to the two possible ways of "currying" a two-argument function $f : \alpha \times \beta \to \gamma$ as one-argument functions:

$$\lambda u. \lambda v. f(u, v) : \alpha \to (\beta \to \gamma)$$

$$\lambda v.\lambda u.f(u,v) : \beta \to (\alpha \to \gamma)$$

We select *a manager* to take narrower scope by using universal instantiation and transitivity of implication to combine the first form with **a-manager** to yield

appointed-a-manager: $\quad \forall X. \; g_\sigma \leadsto X$
$$\multimap f_\sigma \leadsto_t a(z, manager(z), appoint(X, z))$$

We have thus the following derivation

> **every-candidate** ⊗ **appointed** ⊗ **a-manager**
> ⊢ $\quad$ **every-candidate** ⊗ **appointed-a-manager**
> ⊢ $\quad f_\sigma \leadsto_t every(w, candidate(w), a(z, manager(z), appoint(w, z)))$

of the $\forall\exists$ reading of (19).

Alternatively, we could have chosen *every candidate* to take narrow scope, by combining the second equivalent form of **appointed** with **every-candidate** to produce:

> **every-candidate-appointed**:
> $\quad \forall Y. \; h_\sigma \leadsto Y \multimap f_\sigma \leadsto_t every(w, candidate(w), appoint(w, Y))$

This gives the derivation

> **every-candidate** ⊗ **appointed** ⊗ **a-manager**
> ⊢ $\quad$ **every-candidate-appointed** ⊗ **a-manager**
> ⊢ $\quad f_\sigma \leadsto_t a(z, manager(z), every(w, candidate(w), appoint(w, z)))$

for the $\exists\forall$ reading. These are the only two possible outcomes of the derivation of a meaning for (19), as required. We have used our implementation to verify that no other outcomes are possible, since manual verification would be rather laborious.

3.4 Constraints on quantifier scoping

Sentence (24) contains two quantifiers and therefore might be expected to show a two-way ambiguity analogous to the one described in the previous section:

(24) Every candidate appointed an admirer of his.

However, no such ambiguity is found if the pronoun *his* is taken to corefer with the subject *every candidate*. In this case, only one reading is available, in which *an admirer of his* takes narrow scope. Intuitively, this noun phrase may not take wider scope than the quantifier *every candidate*, on which its restriction depends.

As we will soon see, the lack of a wide scope *a* reading follows automatically from our formulation of the semantic contributions of quantifiers without further stipulation. In Pereira's earlier work on deductive interpretation (Pereira 1990, 1991), the same result was achieved through constraints on the relative scopes of glue-level universal quantifiers representing the dependencies between meanings of clauses and the meanings of their arguments. Here, although universal quantifiers are used to support the extraction of properties representing the meanings of the restriction and scope (the variables R and S in the determiner lexical entries), the blocking of the unwanted reading follows from the propositional structure of the glue formulas, specifically the nested linear implications. This is more satisfactory, since it does not reduce the problem of proper quantifier scoping in the object language to the same problem in the metalanguage.

The lexical entry for *admirer* is:

(25) *admirer* N $(\uparrow \text{PRED}) = \text{'admirer'}$
$$\forall X, Y.\ (\uparrow_\sigma \text{VAR}) \rightsquigarrow X \otimes (\uparrow \text{OBL}_{\text{OF}})_\sigma \rightsquigarrow Y$$
$$\multimap (\uparrow_\sigma \text{RESTR}) \rightsquigarrow admirer(X, Y)$$

Here, *admirer* is a relational noun taking as its oblique argument a phrase with prepositional marker *of*, as indicated in the f-structure by the attribute OBL_{OF}. The semantic contribution for a relational noun has, as expected, the same propositional form as the binary relation type $e \times e \to t$: one argument is the admirer, and the other argument is the admiree.

We assume that the semantic projection for the antecedent of the pronoun *his* has been determined by some separate mechanism and recorded as the ANT attribute of the pronoun's semantic projection.[8] The semantic contribution of the pronoun is, then, a formula that consumes the meaning of its antecedent and then reintroduces that meaning, simultaneously assigning it to its own semantic projection:

(26) *his* N $(\uparrow \text{PRED}) = \text{'pro'}$
$$\forall X.\ (\uparrow_\sigma \text{ANT}) \rightsquigarrow X \multimap (\uparrow_\sigma \text{ANT}) \rightsquigarrow X \otimes \uparrow_\sigma \rightsquigarrow X$$

In other words, the semantic contribution of a pronoun copies the meaning X of its antecedent as the meaning of the pronoun itself. Since the left-hand side of the linear implication "consumes" the antecedent meaning, it must be reinstated in the consequent of the implication. The f-structure for example (24) is, then:

---

[8]The determination of appropriate values for ANT requires a more detailed analysis of other linguistic constraints on anaphora resolution, which would need further projections to give information about, for example, discourse relations and salience. Dalrymple (1993) discusses in detail LFG analyses of anaphoric binding.

(27)

$$f: \begin{bmatrix} \text{PRED} & \text{`appointed'} \\ \text{TENSE} & \text{PAST} \\ \text{SUBJ} & g: \begin{bmatrix} \text{SPEC} & \text{`every'} \\ \text{PRED} & \text{`candidate'} \end{bmatrix} \\ \text{OBJ} & h: \begin{bmatrix} \text{SPEC} & \text{`a'} \\ \text{PRED} & \text{`admirer'} \\ \text{OBL}_{\text{OF}} & i: \begin{bmatrix} \text{PRED} & \text{`pro'} \end{bmatrix} \end{bmatrix} \end{bmatrix}$$

with $(i_\sigma \text{ ANT}) = g_\sigma$.

We will begin by illustrating the derivation of the meaning of *an admirer of his*, starting from the following premises:

a: $\qquad \forall H, R, S.\ (\forall x.\ (h_\sigma \text{ VAR}) \rightsquigarrow x \multimap (h_\sigma \text{ RESTR}) \rightsquigarrow Rx)$
$\qquad\qquad \otimes (\forall x.\ h_\sigma \rightsquigarrow x \multimap H \rightsquigarrow Sx)$
$\qquad\qquad \multimap H \rightsquigarrow a(z, Rz, Sz)$

admirer: $\quad \forall Z, X.\ (h_\sigma \text{ VAR}) \rightsquigarrow Z \otimes i_\sigma \rightsquigarrow X$
$\qquad\qquad\qquad \multimap (h_\sigma \text{ RESTR}) \rightsquigarrow admirer(Z, X)$

his: $\qquad\quad \forall X.\ g_\sigma \rightsquigarrow X \multimap g_\sigma \rightsquigarrow X \otimes i_\sigma \rightsquigarrow X$

First, we rewrite **admirer** into the equivalent form

$$\forall X.\ i_\sigma \rightsquigarrow X \multimap (\forall Z.\ (h_\sigma \text{ VAR}) \rightsquigarrow Z \multimap (h_\sigma \text{ RESTR}) \rightsquigarrow admirer(Z, X))$$

We can use this formula to rewrite the second conjunct in the consequent of **his**, yielding

admirer-of-his:
$\qquad \forall X.\ g_\sigma \rightsquigarrow X \multimap$
$\qquad g_\sigma \rightsquigarrow X \otimes$
$\qquad (\forall Z.\ (h_\sigma \text{ VAR}) \rightsquigarrow Z \multimap (h_\sigma \text{ RESTR}) \rightsquigarrow admirer(Z, X))$

In turn, the second conjunct in the consequent of **admirer-of-his** matches the first conjunct in the antecedent of **a** given appropriate variable substitutions, allowing us to derive

an-admirer-of-his:
$\qquad \forall X.\ g_\sigma \rightsquigarrow X \multimap$
$\qquad\qquad g_\sigma \rightsquigarrow X \otimes (\forall H, S.\ (\forall x.\ h_\sigma \rightsquigarrow x \multimap H \rightsquigarrow Sx) \multimap$
$\qquad\qquad\qquad\qquad\qquad H \rightsquigarrow a(z, admirer(z, X), Sz))$

At this point the other formulas available are:

every-candidate:
$\qquad \forall H, S.\ (\forall x.\ g_\sigma \rightsquigarrow x \multimap H \rightsquigarrow Sx)$
$\qquad\qquad \multimap H \rightsquigarrow every(z, candidate(z), Sz)$

appointed:
$\qquad \forall Z, Y.\ g_\sigma \rightsquigarrow Z \otimes h_\sigma \rightsquigarrow Y \multimap f_\sigma \rightsquigarrow appoint(Z, Y)$

We have thus the meanings of the two quantified noun phrases. The antecedent implication of **every-candidate** has an atomic conclusion and hence cannot be satisfied by **an-admirer-of-his**, which has a conjunctive conclusion. Therefore, the only possible move is to combine **appointed** and **an-admirer-of-his**. We do this by first putting **appointed** in the equivalent form

$$\forall Z.\ g_\sigma \rightsquigarrow Z \ \multimap\ (\forall Y.\ h_\sigma \rightsquigarrow Y \ \multimap\ f_\sigma \rightsquigarrow appoint(Z, Y))$$

After universal instantiation of Z with X, this can be used to rewrite the first conjunct in the consequent of **an-admirer-of-his** to derive

$$\forall X.\ g_\sigma \rightsquigarrow X \ \multimap$$
$$(\forall Y.\ h_\sigma \rightsquigarrow Y \ \multimap\ f_\sigma \rightsquigarrow appoint(X, Y)) \otimes$$
$$(\forall H, S.\ (\forall x.\ h_\sigma \rightsquigarrow x \ \multimap\ H \rightsquigarrow Sx) \ \multimap\ H \rightsquigarrow a(z, admirer(z, X), Sz))$$

Universal instantiation of H and S together with modus ponens with the two conjuncts in the consequent as premises yield

$$\forall X.\ g_\sigma \rightsquigarrow X \ \multimap\ f_\sigma \rightsquigarrow_t a(z, admirer(z, X), appoint(X, z))$$

Finally, this formula can be combined with **every-candidate** to give the meaning of the whole sentence:

$$f_\sigma \rightsquigarrow_t every(w, candidate(w), a(z, admirer(z, w), appoint(w, z)))$$

In fact, this is the only derivable conclusion, showing that our analysis blocks those putative scopings in which variables occur outside the scope of their binders.

4 Conclusion

Our approach exploits the f-structure of LFG for syntactic information needed to guide semantic composition, and also exploits the resource-sensitive properties of linear logic to express the semantic composition requirements of natural language. The use of linear logic as the glue language in a deductive semantic framework allows a natural treatment of quantification which automatically gives the right results for quantifier scope ambiguities and interactions with bound anaphora.

The analyses discussed here show that our linear-logic encoding of semantic compositionality captures the interpretation constraints between quantified noun phrases, their scopes and bound anaphora. The same basic facts are also accounted for in other recent treatments of compositionality, in particular categorial analyses with discontinuous constituency connectives (Moortgat 1992a). However, we show elsewhere (Dalrymple et al. 1994) that our approach has advantages over those accounts, in that certain available readings of sentences with intensional verbs and quantified noun phrases that current categorial analyses cannot derive are readily produced in our analysis.

Recently, Oehrle (1993) independently proposed a multidimensional categorial system with types indexed so as to keep track of the syntax-semantic connections that we represent with $\rightsquigarrow$. Using proof net techniques due to Moortgat (1992b) and Roorda (1991), he maps categorial formulas to first-order clauses similar to our semantic contributions, except that the formulas arising from determiners lack the embedded implication. Oehrle's system models quantifier scope ambiguities in a way similar to ours, but it is not clear that it can account correctly for the interactions with anaphora, given the lack of implication embedding in the clausal representation used. A full comparison of the two systems is left for future work.

Appendix

Syntax of the Meaning and Glue Languages

The meaning language is based on Montague's intensional higher-order logic. In fact, in the present paper we just use an extensional fragment with the following syntax:

$$
\begin{array}{llll}
\text{(M-terms)} & M & ::= & c & \text{(Constants)} \\
& & | & x & \text{(Lambda-variables)} \\
& & | & \lambda x\, M & \text{(Abstraction)} \\
& & | & M\, M & \text{(Application)} \\
& & | & X & \text{(Glue-language variables)}
\end{array}
$$

Terms are typed in the usual way; logical connectives such as *every* and *a* are represented by constants of appropriate type. For readability, we will often "uncurry" $M N_1 \cdots N_m$ as $M(N_1, \ldots, N_m)$. Note that we allow variables in the glue language to range over meaning terms.

The glue language refers to three kinds of terms: meaning terms, f-structures, and semantic or σ-structures. f- and σ-structures are feature structures in correspondence (through projections) with constituent structure. Conceptually, feature structures are just functions which, when applied to attributes (a set of constants), return constants or other feature structures. In the following we let A range over some pre-specified set of attributes.

$$
\begin{array}{llll}
\text{(F-terms)} & F & ::= & \uparrow & \text{(Indexical reference)} \\
& & | & f \mid g \mid h \mid \cdots & \text{(F-structure constants)} \\
& & | & (F A) & \text{(Attribute selection)}
\end{array}
$$

$$
\begin{array}{llll}
\text{(σ-terms)} & S & ::= & F_\sigma & \text{(Semantic projection)} \\
& & | & (S A) & \text{(Attribute selection)} \\
& & | & H & \text{(Glue-language variable)}
\end{array}
$$

Glue-language formulas are built up using linear connectives from atomic formulas of the form $S \leadsto_\tau M$, whose intended interpretation is that the meaning associated with σ-structure S is denoted by term M of type τ. We omit the type subscript τ when it can be determined from context.

$$
\begin{array}{llll}
\text{(Glue formulas)} & G & ::= & S \leadsto_\tau M & \text{(Basic assertion)} \\
& & | & G \otimes G & \text{(Multiplicative conjunction)} \\
& & | & G \multimap G & \text{(Linear implication)} \\
& & | & \forall X.\, G & \text{(Quantification over M-terms)} \\
& & | & \forall H.\, G & \text{(Quantification over σ-terms)}
\end{array}
$$

References

Alsina, Alex. 1993. *Predicate Composition: A Theory of Syntactic Function Alternations*. Doctoral dissertation, Stanford University.

Barwise, Jon and Robin Cooper. 1981. Generalized quantifiers and natural language. *Linguistics and Philosophy* 4, 159–219.

Bresnan, Joan and Jonni M. Kanerva. 1989. Locative inversion in Chicheŵa: A case study of factorization in grammar. *Linguistic Inquiry* 20(1), 1–50. Also in E. Wehrli and T. Stowell, eds., Syntax and Semantics 26: Syntax and the Lexicon. New York: Academic Press.

Butt, Miriam. 1993. *The Structure of Complex Predicates*. Doctoral dissertation, Stanford University.

Dalrymple, Mary. 1993. *The Syntax of Anaphoric Binding*. Number 36 in CSLI Lecture Notes. Stanford: Center for the Study of Language and Information.

Dalrymple, Mary, Angie Hinrichs, John Lamping, and Vijay Saraswat. 1993. The resource logic of complex predicate interpretation. In *Proceedings of the 1993 Republic of China Computational Linguistics Conference (RO-CLING)*, Hsitou National Park, Taiwan, September. Computational Linguistics Society of R.O.C.

Dalrymple, Mary, John Lamping, Fernando C. N. Pereira, and Vijay Saraswat. 1994. Intensional verbs without type-raising or lexical ambiguity. In *Proceedings of the Conference on Information-Oriented Approaches to Logic, Language and Computation*, Moraga, California. Saint Mary's College.

Dalrymple, Mary, John Lamping, and Vijay Saraswat. 1993. LFG semantics via constraints. In *Proceedings of the Sixth Meeting of the European ACL*, University of Utrecht, April. European Chapter of the Association for Computational Linguistics.

Fenstad, Jens Erik, Per-Kristian Halvorsen, Tore Langholm, and Johan van Benthem. 1987. *Situations, Language and Logic*. Dordrecht: D. Reidel

Gamut, L. T. F. 1991. *Logic, Language, and Meaning*, volume 2: Intensional Logic and Logical Grammar. Chicago: The University of Chicago Press.

Girard, J.-Y. 1987. Linear logic. *Theoretical Computer Science* 45, 1–102.

Halvorsen, Per-Kristian. 1983. Semantics for Lexical-Functional Grammar. *Linguistic Inquiry* 14(4), 567–615.

Halvorsen, Per-Kristian. 1988. Situation Semantics and semantic interpretation in constraint–based grammars. In *Proceedings of the International Conference on Fifth Generation Computer Systems, FGCS-88*, pages 471–478, Tokyo, Japan, November. Also published as CSLI Technical Report CSLI-TR-101, Stanford University, 1987.

Halvorsen, Per-Kristian and Ronald M. Kaplan. 1988. Projections and semantic description in Lexical-Functional Grammar. In *Proceedings of the International Conference on Fifth Generation Computer Systems*, pages 1116–1122, Tokyo, Japan. Institute for New Generation Systems.

Hendriks, Herman. 1993. *Studied Flexibility: Categories and Types in Syntax and Semantics*. ILLC dissertation series 1993—5, University of Amsterdam, Amsterdam, Holland.

Hepple, Mark. 1990. *The Grammar and Processing of Order and Dependency: a Categorial Approach*. Doctoral dissertation, University of Edinburgh.

Howard, W.A. 1980. The formulae-as-types notion of construction. In J.P. Seldin and J.R. Hindley, eds., *To H.B. Curry: Essays on Combinatory Logic, Lambda Calculus and Formalism*, 479–490. London: Academic Press.

Kaplan, Ronald M. 1987. Three seductions of computational psycholinguistics. In Peter Whitelock, Harold Somers, Paul Bennett, Rod Johnson, and Mary McGee Wood, eds., *Linguistic Theory and Computer Applications*, 149–188. London: Academic Press.

Kaplan, Ronald M. and Joan Bresnan. 1982. Lexical-Functional Grammar: A formal system for grammatical representation. In Joan Bresnan, ed., *The Mental Representation of Grammatical Relations*, 173–281. Cambridge, MA: The MIT Press.

Klein, Ewan and Ivan A. Sag. 1985. Type-driven translation. *Linguistics and Philosophy* 8, 163–201.

Lambek, Joachim. 1958. The mathematics of sentence structure. *American Mathematical Monthly* 65, 154–170.

Miller, Dale A. 1990. A logic programming language with lambda abstraction, function variables and simple unification. In Peter Schroeder-Heister, ed., *Extensions of Logic Programming*, Lecture Notes in Artificial Intelligence. Springer-Verlag.

Montague, Richard. 1974. The proper treatment of quantification in ordinary English. In Richmond H. Thomason, ed., *Formal Philosophy*. New Haven: Yale University Press.

Moortgat, Michael. 1988. *Categorial Investigations: Logical and Linguistic Aspects of the Lambek Calculus*. Doctoral dissertation, University of Amsterdam, Amsterdam, The Netherlands,

Moortgat, Michael. 1992a. Generalized quantifiers and discontinuous type constructors. In W. Sijtsma and H. van Horck, eds., *Discontinuous Constituency*. Mouton de Gruyter, Berlin, Germany. To appear.

Moortgat, Michael. 1992b. Labelled deductive systems for categorial theorem proving. In P. Dekker and M. Stokhof, eds., *Proceedings of the Eighth Amsterdam Colloquium*, 403–423. Amsterdam: Institute for Logic, Language and Computation.

Morrill, Glyn. 1990. Intensionality and boundedness. *Linguistics and Philosophy* 13(6), 699–726.

Morrill, Glyn V. 1994. *Type Logical Grammar: Categorial Logic of Signs*. Studies in Linguistics and Philosophy. Kluwer Academic Publishers, Dordrecht, Holland.

Neale, Stephen. 1990. *Descriptions*. Cambridge, MA: The MIT Press.

Oehrle, Richard T. 1993. String-based categorial type systems. Workshop "Structure of Linguistic Inference: Categorial and Unification-Based Approaches," European Summer School in Logic, Language and Information, Lisbon, Portugal.

Pereira, Fernando C. N. 1990. Categorial semantics and scoping. *Computational Linguistics* 16(1), 1–10.

Pereira, Fernando C. N. 1991. Semantic interpretation as higher-order deduction. In Jan van Eijck, ed., *Logics in AI: European Workshop JELIA '90*, 78–96. Amsterdam: Springer-Verlag.

Pollard, Carl and Ivan A. Sag. 1987. *Information-Based Syntax and Semantics, Volume I*. Number 13 in CSLI Lecture Notes. Stanford: Center for the Study of Language and Information.

Pollard, Carl and Ivan A. Sag. 1993. *Head-Driven Phrase Structure Grammar*. Chicago: The University of Chicago Press.

Roorda, Dirk. 1991. *Resource Logics: Proof-theoretical Investigations*. Doctoral dissertation, University of Amsterdam.

Saraswat, Vijay A. 1989. *Concurrent Constraint Programming Languages*. Doctoral dissertation, Carnegie-Mellon University. Reprinted by MIT Press, Doctoral Dissertation Award and Logic Programming Series, 1993.

Saraswat, Vijay A. 1993. A brief introduction to linear concurrent constraint programming. Technical report, Xerox Palo Alto Research Center, April.

Saraswat, Vijay A. and Patrick Lincoln. 1992. Higher-order, linear concurrent constraint programming. Technical report, Xerox Palo Alto Research Center, August.

Scedrov, Andre. 1993. A brief guide to linear logic. In G. Rozenberg and A. Salomaa, eds., *Current Trends in Theoretical Computer Science*, 377–394. World Scientific Publishing Co.

van Benthem, Johan. 1986. Categorial grammar and lambda calculus. In D. Skordev, ed., *Mathematical Logic and its Application*, 39–60. New York: Plenum Press,

van Benthem, Johan. 1988. The Lambek calculus. In Richard T. Oehrle, Emmon Bach, and Deirdre Wheeler, eds., *Categorial Grammars and Natural Language Structures*, 35–68. Dordrecht: D. Reidel.

van Benthem, Johan. 1991. *Language in Action: Categories, Lambdas and Dynamic Logic*. Amsterdam: North-Holland.

3

The Sorites Fallacy and the Context-Dependence of Vague Predicates

Kees van Deemter

1 Introduction

The sorites (i.e., 'stacked reasoning') paradox of vagueness has puzzled philosophers and logicians for a long time. Since it is well-known that vague predicates tend to be context-dependent, and since recent work in the semantics of natural language has revealed much about context-dependent interpretation, it seemed promising to see whether this work sheds new light on the issue of vagueness. Now as it turns out, this is indeed the case. In the present article, a rather strong claim will be defended: that much of what used to be puzzling about the sorites paradox has now seized to be, and that recent work in semantics, properly viewed, already amounts to a solution of the paradox. For starters, here is a representative version of the paradox.

Suppose you are to judge the height of each of a long series of people, looking at them from a distance that makes a difference in height invisible as long as it amounts to less than, say, 1 cm. Of each of them, you are being asked whether they are short or not. The line-up start from the shortest, who is very short, and ends with the tallest,

I am indebted to Frank Veltman for introducing me to the elusive domain of vague predicates when I was an undergraduate at the University of Amsterdam. More recent sources of inspiration on the topic of vagueness include Johan van Benthem, Louis ten Bosch, Jan van Eijck, David Israel, Stanley Peters, Yoav Shoham, and Ed Zalta – as well as the great marketplace of ideas that CSLI has been for me during my stay there from September 1992 until October 1993, which was funded by the Netherlands Organization for Scientific Research (NWO). Thanks are also due to two anonymous reviewers and to the editors of this volume for helpful comments.

Quantifiers, Deduction, and Context
Makoto Kanazawa, Christopher Piñón, and Henriëtte de Swart, editors
Copyright © 1996, CSLI Publications

who is very tall. The difference between subsequent persons is always less than 1cm, and therefore unnoticeable to you.

Now if you decide that the first person (p_0) is short, you must also judge the next one short, since you can perceive no difference between the two. But then, by the same token, the third person (p_2) must be short as well, and so on indefinitely. In particular, this also makes the last person (p_n) short. But, by assumption, p_n is not short, so a contradiction has been derived. Informally,[1]

1. p_0 is short, and p_n **is not short** (Assumptions).
2. If p_0 is short then p_1 is short.
3. Therefore, p_1 is short.
4. If p_1 is short then p_2 is short.
(...)
2n+1. Therefore, p_n **is short.** $\perp$

The reasoning is elementary, and relies only on a long series of applications of Modus Ponens, and on one crucial premiss, which may be called 'inductive' because of its similarity to the key assumption in a proof by mathematical induction:

Inductive Premiss: For all x and y, if x is short and y is indistinguishable from x, then y is short.

Because of their supposed insensitivity to nonperceptible characteristics of their arguments, words such as *short* are sometimes called Perception Predicates. I will assume that it is the use of a Perception Predicate from which the sorites paradox derives its real cogency, since it is these predicates that give rise to the relation of indistinguishability.[2] Many proposals have been made to resolve this paradox, most of them involving radical departures from classical logic. Let me briefly point out what I take to be the requirements for a solution of the paradox. A proper solution should

1. Explain the plausibility of the argument that leads to the paradox, while showing that the argument is nevertheless invalid.
2. Be consistent with what is known about the meaning of vague

---

[1]The sorites argument is introduced informally, since our solution will depend crucially on the possibility for non-equivalent formalizations.

[2]It is true that some versions of the paradox have been put forward in which perceptual limitations play no role and where, consequently, no Perception Predicates exist. Perhaps the clearest example is Wang's paradox, which centers around the notion of a 'small' natural number; some versions of the paradoxes of the bald man, and the stone heap, are also cases in point. I take it that these versions of the paradox are much less cogent than the ones that use perception predicates, because the conditionals in the even-numbered lines of the argument can simply be negated unless they are supported by a more general assumption such as the inductive premiss.

predicates. Preferably, the solution would be backed-up by empirical evidence.

3. Use as few unmotivated departures from classical logic as possible. A departure can only be motivated by an appeal to (2).

I will argue along the following lines: Empirical research in the last ten years has revealed that *context-dependence* is a pervasive phenomenon in natural language that affects expressions of all kinds, including vague ones. In line with this research, we hypothesize that vague predication must always involve at least two arguments: an object judged *and* a set of elements with which this object is being compared. Consequently, vague predication is incomplete without a 'comparison set', in comparison with whose elements the predication must be understood. Nothing is small or large *per se*, but only in comparison with other things — As will become clear later, this idea is not restricted to vague *predicates*. In particular, the relation of indistinguishability will also prove to be dependent on a 'comparison set' for its interpretation. Predicates and relations alike take their comparison sets from the linguistic and nonlinguistic context of the utterance.

In order to deal with the sorites paradox, the argument itself is viewed as a piece of discourse during which a comparison set is built up. As a result, when the context-dependence of the vague expressions in the argument is taken into account, it turns out that the Inductive Premiss of the sorites argument becomes *ambiguous*. What makes sorites arguments plausible is the fact that, in some of its interpretations, the premiss supports the argument, **but** then it fails to be true by virtue of the meaning of the vague predicate; in other versions, it is true by virtue of the meaning of the predicate, **but** then it does not support the argument. As we will see in section 6, this pattern of explanation holds up even in the case of vague expressions that are not normally viewed as context-dependent.

Our approach to the paradox differs from most other accounts that I know of in that it does not propose a drastic change in the underlying logic. In particular, there is no need to assign unusual interpretations to the conditional, the negation, or the universal quantifier. On the other hand, the interpretation of non-logical constants changes, in that they acquire an additional argument place for a comparison set. The question of how far this proposal is removed from classical logic is also taken up in section 6.

Shortness and smallness will be our paradigmatic examples of a vague predicate, but everything that will be said is meant to be applicable to all vague predicates of the kind that, intuitively speaking, 'measure' something. Such predicates have been studied extensively in standard measurement theory, and their properties are well established.[3] The sentence 'x is small' (or 'x is short') will often be abbreviated by the formula $S(x)$. With the comparison set A as an additional argument, we will write $S(x)^A$ to abbreviate 'x is small with respect to A'.

The structure of this paper is the following. Section 2 will briefly review how context sets have come up in recent natural language research. Section 3 applies the idea of contextualization to the crucial inductive premiss of the sorites argument. Section 4 shows how a properly contextualized version of the paradoxical argument breaks down. Up to and including section 4, I will make use of a convenient simplication, by making some highly specific assumptions about what it can mean for an object x to be 'small with respect to a comparison set' A. Section 5 will relinquish that assumption and ask on what more general assumptions about the meaning of $S(x)^A$ the proposed solution depends. Section 6 deals with some of the questions that the present proposal raises. Most proofs of theorems, as well as a language definition, are delegated to a technical Appendix.

2 Context-Dependence in Natural Language

The notion of context has long been recognized as an important concept in linguistics, especially in formal treatments of the semantics of indexical expressions (Montague 1974, Kaplan 1979). Recent years have seen a rapidly increasing number of applications of the notion of context inside as well as outside linguistics.[4] One prominent example is Situation Semantics, which views meaning as a relation between utterance situations (nonlinguistic context, that is) and described situations. An example that is more directly relevant to our endeavour is the tide of so-called dynamic theories of anaphora (Kamp 1981a, Heim 1982, Barwise 1987), which started to explicate anaphora as a primarily contextual phenomenon. According to these theories, an anaphoric pronoun is only appropriate in a context in which a suitable antecedent for it has been introduced. For instance, if the following pieces of discourse con-

---

[3]The vague predicates (e.g., 'small') of standard measurement theory are predicates that correspond with perceptual relations (e.g., 'visibly smaller') that form a *semi-order*. A structure $\langle A, R \rangle$, where R is the perceptual relation (later denoted '$\prec$'), is a semi-order iff (1) $\forall x(\neg x R x)$, (2) $\forall xyzu((xRy \& zRu) \rightarrow (xRu \lor zRy))$, and (3) $\forall xyzu((xRy \& yRz) \rightarrow (xRu \lor uRz))$. Note that a semi-order allows for indistinguishable elements, i.e., elements x, y such that $x \neq y$, $\neg(xRy)$ and $\neg(yRx)$. For an introduction, see Suppes (1963).

[4]A recent attempt at constructing a unifying theory of the notion of context has been made by John McCarthy and others (McCarthy 1987, Buvač and Mason 1993). We will not use such a general perspective on context, concentrating on one that is more convenient as long as only vagueness is at stake.

stitute the beginning of a story, then version (1) is inappropriate in a sense in which version (2) is not.

(1)*(a) Yesterday, I saw it. (b) A dog barked.
(2) (a) Yesterday, I saw a dog. (b) It barked.

One might say that (2a) adds an individual to the context that is then taken up by *it* in (2b). The occurrence of *it* in (1) is not legitimized by previous context. The main task of an anaphoric theory is then to define under what circumstances a would-be antecedent is *accessible* to a given anaphor. This 'contextual' perspective on anaphora is, by now, widely accepted.

This contextual perspective on things anaphoric has gradually been broadened to include more and more kinds of expressions that depend on context for their interpretation. For example, Barbara Partee analysed the context-dependence of implicit arguments of words such as *local* and *contemporary* along 'anaphoric' lines (Partee 1989).[5] Perhaps most relevant for present purposes is a research track that was started off by Westerståhl in Westerståhl (1985), and where generalized quantifiers can be restricted to contextually available context sets. Subsequent work has extended this approach and shown that NPs of all sorts, and expressions of other categories as well, can be context-dependent (Westerståhl 1985, Carter 1987, Van Deemter 1992). Moreover, the principles that govern the accessibility of context sets turned out to bear close resemblance to those governing the availability of antecedents of anaphoric pronouns. Thus, the possibility begins to present itself of a unified account of context-dependent phenomena, of which pronominal anaphora is just a special case.

Context-based analyses of NPs containing *vague adjectives* have also been forthcoming. An early account[6] is the one in Kamp (1975), which was worked out in much more detail by Ewan Klein. Consider the sentence 'Some small elephant runs'. According to Kamp and Klein, the set of elephants functions as a *comparison set* that somehow sets the standard for what counts as small (Klein 1980). This explains, for example, why a *small elephant* is not necessarily a *small animal*, since elephants are, on average, larger than animals in general.

However, this raises the following question: if a vague adjective can be dependent on *local* (i.e., NP-internal) context, then cannot it also be dependent on *global* (i.e., NP-external) context? After all, the situation of the sorites paradox is one in which no local comparison set is available, as in predicative use. Consider a sentence of the form

---

[5]See Condoravdi and Gawron 1994, this volume, for an interesting follow-up.

[6]According to Kamp, the insight that a noun can act as a comparison set for a vague predicate, is 'probably too old to be traced back with precision to its origin' (Kamp 1975, 127).

'x is small'. How is the standard for smallness determined if it is not through local context? The facts suggest that global context comes to the rescue: the standards of smallness are determined by a comparison set that has been built up during a discourse. There are hints of this idea in Kamp (1975), though no formal account.

Global context was taken into account in a series of later proposals for the semantics of vagueness, including Kamp (1981b), Kwast (1981), and Veltman (1987). In this work, context does not play the role of a comparison set, but is only used to enforce certain kinds of coherence conditions. For instance, in Kamp (1981b), a conjoined[7] or universally quantified formula is false if the addition of the formula to the context causes the context to become 'incoherent'. One thing that makes a context incoherent is when it contains elements a and b that are indistinguishable, while there is a vague predicate S that is true of a and false of b. Due to this move, Kamp can let each of a series of implications of the form $S(p_i) \to S(p_{i+1})$ be true, while the universally quantified formula that has these implications as its instantiations may still be false, because its addition to the context would cause the context to become incoherent. Thus, if a suitable domain is chosen, the inductive premiss of the sorites argument becomes false. Yet, all instantiations of the flawed premiss are true, and this explains why the paradox has intuitive appeal. A similar solution was recently proposed by Veltman and Muskens[8] (Veltman 1987).

In the remainder of this paper, an approach will be described that makes use of the notion of global context, modeled as a comparison set. The starting point is that it makes little sense to ask, of a given object, whether it is small or large unless a comparison set is specified either implicitly or explicitly. Thus, a vague predicate such as *small* is modeled as a *relation* between an individual and a comparison set. For the sake of argument, it will be assumed that this relation is a precise (i.e., non-vague) one. Of course, this assumption may be challenged, but what I intend to show is that this assumption does not make it impossible to account for the sorites paradox. (Cf. section 6.)

Context, in the present proposal, is built up during discourse, in the spirit of dynamic theories of meaning: a discourse is parsed from left to right, and individuals are added to the context as a by-product. To attain the full theory of the semantics of vagueness, many details would have to be sorted out. But fortunately, the structure of a sorites

---

[7]Conjoined formulas become relevant when the paradox is reformulated using a long conjunction of implications, instead of one inductive premiss.

[8]Veltman and Muskens' proposal, which makes use of an idea from Goodman and Dummett that will also play an important role in the present proposal, is discussed at some length in van Deemter (1994).

argument makes it an extremely simple kind of discourse: subsequent elements are judged to be small, and then added to the context.[9]

The proposal of this paper makes use of some of the same building blocks as its predecessors, but amounts to a quite different solution of the paradox. At the root of the difference lies the fact that the present proposal seeks to take the connection between vagueness and anaphora seriously. As was stressed in van Deemter (1994), this anaphoric perspective predicts that the interpretation of a vague predication suffers from an intrinsic *ambiguity* problem, since it is sometimes unclear whether a given expression must be interpreted as anaphoric or not. It will be argued that this ambiguity of vague expressions is exploited in the sorites argument, and the paradox will be analysed as a fallacy that is based on the ambiguity of one of its premisses.

3 Context-Dependent versions of the sorites paradox

In the present section, it will be shown that when comparison sets are taken into account, the inductive premiss of the sorites argument becomes ambiguous, and that there exist highly consequential differences between some of its interpretations.

Given our 'contextual' perspective, what are the possible versions of the crucial premiss? There are three clauses in the premiss to consider: the clause in which x is assumed to be small $(S(x))$, the one in which x and y are compared $(x \sim y)$, and the one in which y is concluded to be small $(S(y))$. Let us first look at the clause in which x and y are compared. The relation of being visibly smaller than something will be viewed as primitive, abbreviated as $\prec$ (Inverse: $\succ$).[10]

$$x \prec y \Leftrightarrow x \text{ is visibly smaller than } y.$$

---

[9]It would be worthwhile to study the context-dependency of vague expressions in greater empirical detail. For example, when a complex NP contains a vague adjective, local and global factors interact. Thus, *large natural numbers* can only have a non-empty denotation if the comparison set for *large* is contextually restricted to some finite subset of the natural numbers, since only finitely many natural numbers are smaller than any given number. (Cf. Dummett 1975, p. 303.) A linguistically more interesting case arises when a sentence such as 'Some small elephants run' refers to the elephants in a zoo. One possible comparison set is the entire set of Elephants, and another is Elephants ∩ (Animals-in-the-Zoo). But Animals-in-the-Zoo seems to be a less natural comparison set. Since our sole purpose is to explain the sorites paradox, it suffices to deal with predicative uses of vague adjectives, as in 'x is small', and in such cases global context-dependence is the only thing to worry about.

[10]I interpret Van Benthem's results on the semantics of the vague predicate *late* in van Benthem (1983) as showing that, logically speaking, little hinges on whether this relation is taken as primitive or defined on the basis of the vague predicate.

$x \succ y \Leftrightarrow y$ is visibly smaller than x.

$x \sim y \Leftrightarrow \neg x \prec y \ \& \ \neg y \prec x$.

How can context be relevant for the interpretation of (in)distinguishability? One interesting answer has been suggested by Goodman and examined more closely by Dummett: the elements that a context makes available can be viewed as resources that can help distinguish observed individuals. Thus, one might first define 'x is smaller than y with respect to A' (hence $(x \prec y)^A$) and then 'x is indistinguishable from y with respect to A' (hence $(x \sim y)^A$). We would like to think of $(x \prec y)^A$ as expressing a relation that involves observation assisted by reason. Therefore, it is natural to let $(x \prec y)$ have $(x \prec y)^A$ as a logical implication, and since we do not want to prejudge the question of whether x and y themselves must be elements of A, the clause $x \prec y$ will be included as a separate disjunct:

$(x \prec y)^A \Leftrightarrow x \prec y \ \vee$
$\exists h \in A : (h \prec y \ \& \ \neg h \prec x) \vee (x \prec h \ \& \ \neg y \prec h).$

For example, in the following situation, h helps to tell x and y apart:

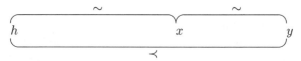

The definition implies that $(x \prec y)^A$. One way to motivate this is by arguing that a perceptible difference between two elements arises if and only if the difference in their sizes is at least as great as some constant value. This constant will now be assumed to be the same for all elements, and called a Just Noticeable Difference (JND) (Suppes and Zinnes 1963). Relativized *in*distinguishability can now be defined as follows:

$(x \sim y)^A \ \Leftrightarrow_{Def} \ \neg (x \prec y)^A \ \& \ \neg (y \prec x)^A.$

In effect, 'indistinguishability with respect to A' means that A provides no help element to tell x and y apart. It is easy to see that $x \sim y \Leftrightarrow (x \sim y)^{\{x\}} \Leftrightarrow (x \sim y)^{\{y\}} \Leftrightarrow (x \sim y)^{\{x,y\}}$. Relativized indistinguishability has many interesting properties such as, for example, transitivity [Appendix, 2]. Also, the Goodman/Dummett construction is stable, in the sense that iteration does not change it [Appendix, 3].[11]

At this point we have a plausible candidate for a relativized version of the clause that says 'x is indistinguishable from y'. So how about smallness, how can that notion be relativized to a comparison set? Let

---

[11] These observations were also made by Frank Veltman and Reinhard Muskens who show that the Goodman/Dummett notion of indistinguishability does not lead to the sorites paradox (Veltman 1987).

us, for now, simplify and assume that there is no ambiguity in the notion of 'Smallness with respect to a comparison set A', and that it involves a comparison between two sets: the set of elements in A that are visibly smaller than the element x that is being judged, and the set of elements in A that are visibly larger than it. One way to implement this idea which is simple enough to work with while also being consistent with much of what other authors have proposed[12] is expressed in the following meaning postulate:

Comparison Postulate: $S(x)^A \Leftrightarrow |\{y \in A : y \succ x\}|$ is greater than $|\{y \in A : x \succ y\}|$.

The following picture shows which elements of A the Comparison Postulate makes responsible for x's being small or not. $G[x]^A$ abbreviates $\{y \in A : y \succ x\}$ and $K[x]^A$ abbreviates $\{y \in A : x \succ y\}$:

The elements in between $K[x]^A$ and x are not taken into account, and neither are those in between x and $G[x]^A$. As will become clear later, not much hinges on the specifics of the Comparison Postulate (see section 5 and also Appendix, 9). But even if the Comparison Postulate is taken for granted, there are plenty of ways to render the inductive premiss of the sorites argument. As with anaphora, there is a resolution problem: each of the three clauses can, in principle, be relativized to any context set. The following options seem reasonable: each of the three clauses may be interpreted (a) with no context set at all, (b) with respect to some context set A that is given at the beginning of the argument, or with respect to (c) $A\cup\{x\}$, (d) $A\cup\{y\}$, and (e) $A\cup\{x,y\}$. As a result, there are 125 different versions of the sorites paradox.

Getting to grips with all these versions of the paradox may seem a daunting task, but their number can be reduced. Firstly, let us, in accordance with a dynamic perspective on interpretation, assume that elements in a context set are always only assembled from left to right. Consequently, the only three options for the *first* clause, $S(x)$, are (a), (b), or (c). The element y simply hasn't come up yet.

Secondly, let us assume that the formula $x \sim y$ is symmetric, in that it either introduces both x and y into the comparison set, or none of the two. So once more, only three options are left: options (a), (b), and (e). Unfortunately, these two assumptions do not reduce the options for the third clause, $S(y)$, however, so we are still left with 45 versions.

---

[12]Compare, for example, Klein's treatment of attributive *small* (Klein 1981), or Dummett's treatment of the word *small*, applied to natural numbers (Dummett 1975, p. 303).

However, this number can be reduced further. It is easy to see that all of $(x \sim y)^A$, $(x \sim y)^{A \cup \{x\}}$, $(x \sim y)^{A \cup \{y\}}$, and $(x \sim y)^{A \cup \{x,y\}}$ are equivalent, and this reduces the number of options for the indistinguishability clause to two: plain $\sim$ and '$\sim$ with respect to A'. To further reduce the number of options for the other two clauses, discard option (a) for the clauses $S(x)$ and $S(y)$, since S requires that some comparison set be specified. At this point, we are left with the following options, in which x, y and A are universally quantified. Indicated are the sets to which the respective clauses may be relativized. A dash indicates the option of using no context set at all.

$(S(x)$	&	$x \sim y)$	$\rightarrow$	$S(y)$
A		$-$		A
$A \cup \{x\}$		A		$A \cup \{x\}$
				$A \cup \{y\}$
				$A \cup \{x,y\}$

But, it is easy to see that even most of these options are logically equivalent if 'smallness with respect to a comparison set' is interpreted as in the Comparison Postulate. In particular, it will never make any difference whether the comparison set for a certain clause equals A, $A \cup \{x\}$, $A \cup \{y\}$, or $A \cup \{x,y\}$, given that x and y are perceptually indistinguishable. In other words: options (b), (c), (d) and (e) are logically identical in the case of the two S-clauses. Consequently, only one version of the two S-clauses needs to be distinguished, and only two versions of the second clause. For reasons of general plausibility in a dynamic setting, we will select a version in which $S(x)$ is evaluated with respect to A, while $S(y)$ is evaluated with respect to $A \cup \{x\}$.[13]

$(S(x)$	&	$x \sim y)$	$\rightarrow$	$S(y)$
A		$-$		$A \cup \{x\}$
A		A		$A \cup \{x\}$

Neither of these two versions leads to paradox. The first one can easily be falsified. Let z be the only element of the comparison set A, and assume x can just barely be discerned to be smaller than z. In other words, the difference in size between them equals a JND. Consequently,

---

[13]Note that the equivalence between the different relativized forms of the two S-clauses depends on the fact that they occur in a context in which $x \sim y$. In other contexts, where $x \sim y$ does not hold, there are important differences between $S(x)^A$ and $S(x)^{A \cup \{y\}}$.

if $x \sim y$, and y is larger than x, then y is *not* distinguishably smaller than z. Under these circumstances, $G[x]^A = \{z\}$ and $K[x]^A = \emptyset$, while $G[y]^A = K[y]^A = \emptyset$, and consequently $S(x)^A$, while not $S(y)^A$.

The second version of the premiss cannot be falsified. Such a version of the premiss, which must be true due to the meaning of the terms in the premiss, will be called *valid*. On the other hand, this version does not support the sorites argument. To see that it is valid, assume $S(x)^A$ while not $S(y)^A$. Given the Comparison Postulate, this can only be true if some element h of A behaves differently towards x and y, in other words, if $h \prec y$ while not $h \prec x$, or if $x \prec h$ while not $y \prec h$. But under such circumstances, h fulfils the function of a help element in the sense of Goodman/Dummett, contradicting $(x \sim y)^A$. So, this version of the premiss is valid. To see that it does not support the paradox, let Size be a function that maps individuals to their (real-valued) size in the relevant dimension. Let x_0 be an element that is judged to be small with respect to the comparison set A_0. Then $\text{Size}(x_0) +$ JND functions as an upper limit on the sizes of elements that can be inferred, by the inductive premiss, to be small. The argument goes

$S(x_0)^{A_0}$ and $(x_0 \sim x_1)^{A_0}$, therefore
$S(x_1)^{A_0 \cup \{x_0\}}$; but $(x_1 \sim x_2)^{A_0 \cup \{x_0\}}$, therefore
$S(x_2)^{A_0 \cup \{x_0, x_1\}}$; *etcetera*.

Note that x_0 is an element of all comparison sets after the initial one. As soon as one arrives at x_i such that $(x_0 \sim x_{i-1})$ and $(x_0 \prec x_i)$, then $(x_{i-1} \prec x_i)^{A \cup \{x_0, x_1, x_2, \ldots x_{i-2}\}}$. At that point, x_i is not indistinguishable from its predecessor (with respect to the relevant comparison set) and the argument comes to a halt. A more detailed illustration of how this can happen is offered in section 4.

To sum up, when comparison sets are brought to bear on vague predicates, the result is that the inductive premiss of the sorites argument displays an 'anaphoric' ambiguity, that hinges on the choice of comparison sets for the respective clauses of the premiss. The premiss becomes ambiguous between two sorts of interpretations. One is strong enough to support the sorites argument but is invalid; the other is too weak to support the paradox, but happens to be valid. I believe that this ambiguity is a plausible cause of confusion on the part of language users.[14]

At this point, the *invalid* version of the inductive premiss need not occupy us much. The *valid* version, however, which employs

---

[14]Note that even the invalid version of the premiss could be considered 'almost' valid, in the following precise sense: for any x, y for which $G[x]^A = G[y]^A$ and $K[x]^A = K[y]^A$, if $S(x)^A$ and $(x \sim y)^A$, then also $S(y)^A$. A stronger, bidirectional version of this proposition is proved in Appendix, 4.

Goodman/Dummett-style indistinguishability calls for some illustration.

4 Illustration: How the paradox breaks down

To illustrate how comparison sets cause the version of the sorites argument that makes use of the Comparison Postulate for smallness and the Goodman/Dummett notion of indistinguishability to break down, let us go through an actual sorites argument. We are dealing with the following version of the inductive premiss:

Premiss **P**: $\forall y \forall y \forall A \, ((S(x)^A \,\&\, (x \sim y)^A) \rightarrow S(y)^{A \cup \{x\}})$.

From now on, I will not speak of the sizes of human beings, since we all have a long history of acquaintances with people, while it is not clear how all of these affect our standards of comparison. (Memory issues become prevalent, etc.) Instead, let us talk of individuals, Martians, say, none of which the observer has ever seen before, and all of which have a height between 0.60 m and 1 m. Other dimensions than vertical size are disregarded. Assume a line-up consists of individuals that are just 1 mm apart, starting at 0.60 m. Let us number the Martians in the line-up with increasing length, so x_0 has length 0.60, and x_i has length 0.60 $+i$ mm. A Just Noticeable Difference (JND) equals 10 mm. We will assume that the initial standard of comparison is something like one's own height. In my case this is sufficiently taller than 1 m, so the first Martian (named x_0) will be called small.

Using the assumptions that were just made, the following is predicted, where the letter I denotes the speaker:

0. $S(x_0)^A$, since $A = \{I\}$, while $G[x_0]^A = \{I\}$ and $K[x_0]^A = \emptyset$
1. $S(x_1)^A$, since $A = \{I, x_0\}$, while $G[x_1]^A = \{I\}$ and $K[x_1]^A = \emptyset$
....

9. $S(x_9)^A$, since $A = \{I, x_0, .., x_8\}$, and $G[x_9]^A = \{I\}$ and $K[x_9]^A = \emptyset$
10. $\neg S(x_{10})^A$, since $A = \{I, x_0, .., x_9\}$, and $G[x_{10}]^A = \{I\}$ and $K[x_{10}]^A = \{x_0\}$.

Somehow, the sorites argument must break down at (or before) x_{10}. There are prospective applications of **P** from (0) to (1), from (1) to (2),..., and from (9) to (10), with different instantiations of x, y, A. For example, one gets from (0) to (1) by chosing $x := x_0$, $y := x_1$, and $A := \{I\}$, observing that $(x_0 \sim x_1)^{\{I\}}$, since both are distinguishably smaller than the speaker. This works until one arrives at step (10). By assumption, x_{10} is 1 mm taller than x_9, making it, for the first time, distinguishably taller than the initial element x_0. As a result, $\neg((x_9 \sim x_{10})^{\{I, x_0, \ldots, x_8\}})$, since $(x_0 \sim x_9)$, but $(x_0 \prec x_{10})$. In other words: after having inferred that x_9 must be small, relative to the context at hand, there happens to be no suitable object around to instantiate the antecedent of **P**. Of course such an object might have been around, for example, an element $x_{10'}$ that is just 0.5 mm taller than x_9, and then another after $x_{10'}$, and so on. However, in that way one will always only infer the smallness of elements that are smaller than x_{10}, and no paradoxical conclusion is inferred. The value Size(x_0) + JND functions as a limit: only x that are smaller than this value (i.e., smaller than x_{10}) are Goodman/Dummett-style indistinguishable from x_9.

Observe that, in this example, elements stop being called small at precisely the same time when they stop being Goodman/Dummett-style indistinguishable from x_9. Should our initial comparion set have consisted of two elements, I and an element I' that is just 1 mm taller than I, then x_{10} would also have been Goodman/Dummett distinguishable from x_9, but it would still have been small, since $G[x_{10}]^A = \{I, I'\}$ and $K[x_{10}]^A = \{x_0\}$.

Starting from this example, it is easily seen that the paradox can break down in one of the following three ways: (i) the case illustrated above, in which the argument breaks down through an element x_i that has the same $G[x_i]^A$ as its predecessor, but an enlarged $K[x_i]^A$; (ii) the same $K[x_i]^A$, but a diminished $G[x_i]^A$; or (iii) through an element that is located in such a way that $G[x_i]^A$ and $K[x_i]^A$ change at the same time. The second of the three situations is depicted in the following, where x_0 is absent, so I becomes the element that causes the breakdown:

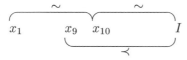

The last of the three situations is depicted in the following:

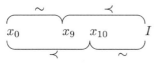

So, why is the paradox flawed? Well, firstly, it was assumed that a vague notion such as *small* requires a comparison set. Now *whatever* finite set one takes as the initial comparison set A_0, the result is some finite 'norm', implying some upper limit on what is small with respect to that norm. Of course, the norm will change when newly-judged individuals enter the comparison set. But, as long as the elements judged are *small* with respect to the current comparison set, they will only cause the norm to go *down*. Adding a small element to a comparison set may sometimes cause a small element to become not-small, but it can never cause a non-small element to become small:

If $S(y)^A$, then $\|S^{A\cup\{y\}}\| \subseteq \|S^A\|$.[15]

As a result, all the individuals $(x_0, x_1, ...)$ that are added in the course of a sorites argument first leave the norm of the initial comparison set intact, and in the somewhat longer run, they cause the norm to become even stricter, causing an even earlier break-down of the argument.

Before moving on to a perspective in which the Goodman/Dummett construction looses its privileged position, let me point out that many interesting variations on the theme of this construction can be conjured up. For example, it may be objected (as was already pointed out in Dummett 1975) that using arbitrary help-elements to make distinctions predicts *unlimited resolution*. Take any two elements x and y that have different sizes, and take any perceiver, no matter how imperfect. Then Goodman/Dummett predict that x and y can be told apart by this perceiver provided an appropriately sized help-element can be found. This objection can be countered by replacing the Goodman/Dummett construction by a more subtle one. For example, the class of help-elements can be limited to elements that have occurred in the recent past.[16] Also, a stochastic aspect may be added, to account for the fact that an observer is never completely consistent in his or her judgments. For instance, it may be required of a help-element that it is judged as greater than x on at least $m\%$ of judgments, while it is judged as greater than y on less than $n\%$ of judgments, where $n \leq m$. For example, with $n = m = 90$,

$(x \prec y)^A_{90\%} \Leftrightarrow \exists h :$ on 90% of measuring events, h is judged greater than x, while on less than 90% of judgments h is judged greater than y, or similarly for help elements that are *smaller* than y,

where $(x \sim y)^A_{90\%}$ is defined as $\neg(x \prec y)^A_{90\%}$ & $\neg(x \prec y)^A_{90\%}$. This would do away with unlimited resolution, while it would still lead to a valid inductive premiss if it is combined with the following, stochastic version of smallness:

---

[15]The extension of *small* with respect to comparison set A is denoted $\|S^A\|$. The formula in the text is the later constraint S_2. A strengthening, saying that smallness becomes extensionally *smaller* when sufficiently many small elements are added can be found in Appendix, 7.

[16]The role of help-elements plus the existence of memory limitations could, for example, explain the reported fact that frogs (i.e., cold-blooded animals) sometimes fail to jump out of a slowly heated water container, even when the water is heated to reach boiling point. (Jan van Eijck, personal communication.)

$S_{90\%}(x)^A \Leftrightarrow_{Def} |G_{90\%}[x]^A| > |K_{90\%}[x]^A|$, where $G_{90\%}[x]^A =_{Def}$ $\{z \in A : z \succ x$ on at least 90% of size judgments$\}$, and $K_{90\%}[x]^A$ $=_{Def} \{z \in A : z \prec x$ on at least 90% of size judgments$\}$.

5   Generalization to other versions of the paradox

The previous sections have shown how the sorites paradox can be explained away, provided some rather strong assumptions are made as to how context affects the key concepts of the paradox. The two assumptions used were that smallness is defined as in the Comparison Postulate, while indistinguishability is defined as either perceptual indistinguishability, or Goodman/Dummett-style indistinguishability. The results of these earlier sections are pleasing to the extent to which these assumptions are considered realistic. But there are uses of vague predicates that do not satisfy them, and these may also give rise to a sorites argument. For instance, one might understand smallness with respect to a comparison set $S(x)^A$ to mean the following:

Modified Comparison Postulate: $S(x)^A \Leftrightarrow_{Def}$ Size(x) is distinguishably smaller than Size(y), for a majority of elements $y \in A$.

which is much like the definition in the Comparison Postulate, except that it compares the elements that are distinguishably greater than x with *all* the other elements of A, rather than only with those that are distinguishably smaller. Deviating more drastically from earlier assumptions, one might understand smallness as involving a comparison between the size of x and the average of the absolute sizes of elements in A:

Average Size postulate: $S(x)^A \Leftrightarrow_{Def}$ Size(x) is smaller than the average of the sizes of elements of A.

In addition to such alternatives for the notion of smallness, the notion of *indistinguishability* can be understood in various ways as well: as exact equality, as Goodman/Dummett-style indistinguishability, as perceptual indistinguishability, or as some even more relaxed notion of similarity, for example. Now given all this latitude, can the sorites paradox be explained away in all its versions?

For simplicity, let us once more fix the 'anaphoric' ambiguities involved in the determination of the comparison sets for the different clauses of the inductive premiss, and assume that the formula **P** captures its appropriate form:

Premiss **P**: $\forall x \forall y \forall A \left((S(x)^A \ \& \ (x \sim y)^A) \rightarrow S(y)^{A \cup \{x\}} \right)$.

But, whereas **P** in earlier sections made use of S and $\sim$ as *defined* notions, this time they are open to interpretation. For example, $(x \sim y)^A$ may be defined in any reasonable way, including ways that make the relativization to A inessential.[17] As a result, $\mathbf{P}$ does not prejudge the issue of what smallness and indistinguishability mean.

My strategy will be the following. First, the notion of a *version* of $\mathbf{P}$ (for short: $\mathbf{P}$-version) will be made more precise and then a formulation will be proposed of what it means for a $\mathbf{P}$-version to *support* the sorites paradox. Next, it will be shown that, under some highly plausible assumptions, no $\mathbf{P}$-version can be valid and at the same time support sorites. At that point, it has been shown that any version of sorites must be flawed. To then explain its appearance of validity, I will point to large classes of $\mathbf{P}$-versions that are valid and to large classes that support sorites. Like in the specialized case that we have been dealing with in previous sections, it is this duality that explains the appearance of validity.

'Supporting sorites' defined. A $\mathbf{P}$-version can be viewed as a pair $\langle P, MP \rangle$, where MP is a set of Meaning Postulates governing the use of $\sim$ and S. The set MP may or may not amount to a full definition of $\sim$ and S. For example, MP may consist of just the Goodman/Dummett clause.

The definition of what it means for a $\mathbf{P}$-version to support sorites must reflect what is essential about the sorites argument. Informally speaking (i.e., without taking comparison classes into account), we can take this essential quality to be that elements of arbitrarily great size are inferred to be small. This, of course, leads to paradox if you accept that there are non-small objects. — Formally, the essential sorites property can be rendered as the existence of *unlimited S-sequences with finite initial comparison set*: In this formalization, we will continue to conflate individuals and *names* of individuals, using x_0, x_1, etc. for both ends.

An **Unlimited S-sequence** in a model M is a sequence of individuals $x_0, x_1, ..., x_i, ...$ in M's domain, such that $S(x_0)^{A_0}$ & $S(x_1)^{A_0 \cup \{x_0\}}$ & ... & $S(x_i)^{A_0 \cup \{x_0, ..., x_{i-1}\}}$ & ... and such that for all sizes m, there exists n in the sequence such that $\text{Size}(x_n) \geq m$.

The set A_0 is called the initial comparison set of the sequence. 'Supporting sorites' can now be defined as leading to unlimited S-sequences

---

[17]The relativization to A is inessential if $(x \sim y)^A$ is equivalent to a clause $\phi(x, y)$ in which A does not occur. For example, this is the case if $(x \sim y)^A$ is defined as $x \sim y$.

with a finite[18] initial comparison set in all models that are sufficiently dense and that contain arbitrarily large elements:

> A **P**-version $\langle P, MP \rangle$ *supports sorites* $\Leftrightarrow_{Def}$ For any model $M = \langle D, I \rangle$ that verifies all postulates in MP, and for any $A_0 \subseteq D$, if $M \models S(a)^{A_0}$, for some element a, and D is densely ordered by $<$ and contains elements of arbitrarily large size and $M \models$ **P**, *then* there exists an unlimited S-sequence in M starting with $S(a)^{A_0}$, where A_0 is finite.

Density may be defined in any of several ways. A suitable definition that does not itself depend on the notion of smallness or on the notion of indistinguishability is the one that is customary in logic:

$$\forall x, y \in D(x < y \rightarrow \exists z(x < z \,\&\, z < y)).$$

No valid P-version supports sorites. It is intuitively plausible that it is impossible for arbitrarily large objects to be small. But since comparison sets complicate the story, a formal proof is required. It must be proven that the existence of unlimited S-sequences with finite A_0 is at odds with the meaning of *small*. One way to make this point is by formulating constraints on what *small* (i.e., S) can mean:[19]

Constraint S_1. In finite contexts, something can only be small if something else is not small:

> For every finite comparison set A, if $\exists x S(x)^A$, then $\exists y \neg S(y)^A$.

Constraint S_2. Adding an element y that is small with respect to a comparison set to that comparison set will never add any elements to the extension of *small*:

> If $S(y)^A$, then $\|S^{A \cup \{y\}}\| \subseteq \|S^A\|$.

Constraint S_3. Elements y that are larger than a given element x which is not small with respect to a comparison set are not small with respect to that comparison set either:

> If $\neg S(x)^A$ and $x < y$ then $\neg S(y)^A$.

It is easily proven that all three families of notions of smallness that were mentioned at the beginning of this section fulfil $S_1 - S_3$. (See Appendix, 8.) Together, $S_1 - S_3$ imply that there can be no unlim-

---

[18] The reason for the finiteness condition is that, in an infinite setting, unlimited S-sequences are not paradoxical. For example, suppose $A_0 =$ the set of natural numbers. Let S be defined as in the Comparison Postulate, where $\prec$ is the ordinary 'smaller than' relation between natural numbers. Then $\forall x_i (S(x_i)^{A_0 \cup \{x_0, \dots, x_{i-1}\}})$, since $A_0 \cup \{x_0, \dots, x_{i-1}\}$ contains a majority of elements that are greater than x_i. Consequently, the unlimited S-sequence does not lead to paradox. (Cf. Dummett 1975, p. 303, for support of this claim.)

[19] Compare also the constraints proposed in van Benthem (1983).

ited S-sequences that are based on initial comparison sets with finite cardinality. (See Appendix, 5.). But then it follows that supportive **P**-versions themselves, which will in some domains lead to unlimited S-sequences, are at odds with a proper understanding of *small*. Consequently, no **P**-version can support sorites and be valid at the same time.

Existence of valid P-versions, and of supportive P-versions. This leaves us with the task of showing that there is an ambiguity between interpretations of **P** that are valid and interpretations that support sorites. In sections 3 and 4, this was shown to be true under certain strong assumptions about the meaning of the vague predicate and the notion of indistinguishability. To get a handle on sorites arguments that deviate from these assumptions, we need some further results. Firstly, it is possible to give a more manageable characterization of the class of **P**-versions that support sorites, by taking into account that the notion of smallness is irrelevant here. What matters is only whether the notion of indistinguishability employed in **P** is 'loose enough'. More precisely, and taking comparison sets into account, the crucial question is whether there exist **unlimited ∼- paths**:

An **Unlimited ∼-path** in M is a sequence $x_0, x_1, \ldots$ in M's domain such that, for some initial comparison set A_0, it holds that $(x_0 \sim x_1)^{A_0}$ & $(x_1 \sim x_2)^{A_0 \cup \{x_0\}}$ & ..... & $(x_{i-1} \sim x_i)^{A_0 \cup \{x_0, \ldots, x_{i-2}\}}$ & ..., and such that, for any size m, an element x_n can be found in the sequence such that $\mathrm{Size}(x_n) \geq m$.

An example is the ordinary notion of perceptual indistinguishability. It is easy to see that **P** can be used to infer the smallness of arbitrarily tall persons whenever there exist unlimited ∼-paths. (See Appendix, 6.) Thus, there are plenty of **P**-versions that support sorites.

Now how about validity? Which interpretations of the premiss validate it? Here are some simple results. First, let us formulate a tentative new condition on smallness:

Constraint S_4:
If $\mathrm{Size}(x) = \mathrm{Size}(y)$, then $S(x)^A \to S(x)^{A \cup \{y\}}$.

S_4 is less cogent than the earlier $S_1 - S_3$,[20] but it is useful to prove such theorems as the following:

Theorem i: Assume indistinguishability implies exact equality of sizes,[21] while $S(x)^A$ satisfies S_4. Then any definition of $S(x)^A$ validates **P**.
Proof: Immediate.

---

[20] Observe, for example, that the Modified Comparison Postulate contradicts S_4.
[21] For example, this holds if $(x \sim y)^A \Leftrightarrow_{Def} \mathrm{Size}(x) = \mathrm{Size}(y)$.

An analogous result may be obtained for a notion of indistinguishability that contains a 'safety valve' that explicitly forbids to call two elements indistinguishable if one of them is small and the other is not: Let S be any notion of smallness and let *Indist* be any notion of indistinguishability. Define $(x \sim y)^A \Leftrightarrow_{Def} x$ *Indist* y & $(S(x)^A \leftrightarrow S(y)^A)$. Then, obviously, $(S(x)^A$ & $(x \sim y)^A) \to S(y)^A$ is valid. The safety valve may appear artificial, but it hinges on a difference between x and y that seems pre-eminently relevant to their (in)distinguishability, namely whether they differ in terms of smallness. Be this as it may, it seems possible that indistinguishability is confused with equality, with equality of size, with 'indistinguishability plus safety valve', or with some other notion that would cause **P** to be valid.[22].

We have just seen that there are other notions of indistinguishability than that of Goodman and Dummett, that still validate **P**, no matter how *small* is defined. Conversely, one may stick to the Goodman/Dummett variety and look for other definitions of *small* that validate **P**. Thus, one may generalize an earlier result concerning the Goodman/Dummett notion of indistinguishability:

> **Theorem ii:** Assume indistinguishability is defined as by Goodman/Dummett. Then **P** is valid iff for all x and A, $S(x)^A \Leftrightarrow \Theta(G[x]^A, K[x]^A)$, for some relation Θ between sets.
> *Proof:* Appendix, 9.

Simple examples of Θ are $\lambda X \lambda Y(|X| > |Y|)$ (as in the Comparison Postulate), $\lambda X \lambda Y(|X| > 2.|Y|)$, or $\lambda X \lambda Y(a \in X)$. Again, the resulting versions of **P** obviously do not support the paradox, since they employ Goodman/Dummett-style indistinguishability.

Perhaps the best way to sum up the spirit of the present section is by putting forward the following hypothesis:

> **Hypothesis:** For any plausible notion of smallness, there exists a plausible notion of indistinguishability that validates **P**. Conversely, for any plausible notion of indistinguishability, there exists a plausible notion of smallness that validates **P**.

Proof of this hypothesis must await a more complete characterization of what it means for notions of indistinguishability and smallness to be plausible.

---

[22]Most of the notions of indistinguishability that were just discussed lead to non-supportive **P**-versions, since they do not allow unlimited $\sim$-paths. This is also true for the version involving a safety valve, provided there are x and A such that $S(x)^A$.

6 Conclusion: Vagueness and Context-Dependence

In the above, it was shown how the assumption that vague predications involve an explicit or implicit comparison set bears on the sorites paradox, through ambiguities in what constitutes the comparison set for the different clauses of the inductive premiss of the sorites argument. In the introduction of this paper, logical conservativity was defended to be an important *desideratum* for a solution to the sorites paradox. So, in how far can our solution be regarded as logically conservative?

It has been noted that the only non-classical element in the proposal consists of relativizations of atomic clauses to comparison sets. Thus, a mechanism that is often only implicit in natural languages was made explicit in the logical object language. An alternative approach is also possible, which leaves comparison sets implicit. Mimicking the work of Groenendijk and Stokhof on Dynamic Predicate Logic (Groenendijk and Stokhof 1991), one might use an object-language that is syntactically equal to a variant of predicate logic, while its semantics ensures that comparison sets are provided. One might think of interpretive rules such as the following (where M is a model with universe E and interpretation function I, while $A \subseteq E$ is a comparison set):

$$M, A \models S(a)\&S(b) \Leftrightarrow$$
$$M, A \models S(a) \text{ and } M, A \cup \{I(a)\} \models S(b),$$

where the individual denoted by a enters the comparison set once a has been interpreted. Thus, interpretation would become truly dynamic. Given a particular dynamic semantics of this kind, it would be natural to ask for a deductive system that is sound and complete with respect to it.

The main reason not to 'go implicit' is that it would have forced us to rule out any ambiguities in the assignment of comparison sets, and ambiguity is a key element in the explanation of the paradox. Less crucially, it would also have tended to over-simplify the way in which comparison sets grow. For example, elements are not only added to a comparison set, but also substracted, since memories of the elements of comparison sets fade away and become irrelevant.

Our choice to keep references to context explicit has as a consequence that the logic is monotonic in the following sense: If premises are added to the premises of a valid argument, then a valid argument must again result. If $\Gamma \models S(x)^A$, then also $\Gamma \cup \{\phi\} \models S(x)^A$, no matter what ϕ is. The language of relativizations can be embedded in classical logic by writing $S(x)^A$ as a two-place relation: $S(x, A)$. Of course, this language is used to model a non-monotonic situation. After all, a natural language sentence of the form 'x is small' can be true at a given point in a discourse (e.g., as a conclusion from premisses Γ), but false at some later point (e.g., as a conclusion from premisses $\Gamma \cup \{\phi\}$). Non-monotonicity is delegated to the translation from natural language to logic. If 'x is small' is uttered in a context in which A is the comparison set for smallness, it is represented by means of a formula in which A appears; in a context in which a superset A' of A is the comparison set, A is replaced by A'.

This paper has advocated an approach to the sorites paradox of vagueness in which the notion of context-dependence plays a central role. This raises the question of whether vagueness must always involve context-dependence and how possible counterexamples affect the present proposal. It is at least conceivable that there exist vague predicates that are not context-dependent. For example, their vagueness may result from perceptual or judgmental limitations, while the standards for comparison are invariable. A tentative example would be the vague predicate *healthy*. Arguably, whether a person is healthy is not dependent on the context in which the person's health is considered, but only on the requirements of survival and procreation. Let us define an artificial predicate *smoll*, which is similar to the adjective *small*, except that it is 'absolute', like *healthy*:

$$Smoll(x) \Leftrightarrow_{Def} x \prec I.$$

In other words, something is *smoll* if and only if it is distinguishably smaller than the speaker. Using the definiens of *smoll*, an informal version of the inductive sorites premiss says

Inductive Premiss: For all x and y, if $x \prec I$ and y is indistinguishable from x, then $y \prec I$.

Let us define, as before, the contextualized version $(x \sim y)^A$ in Goodman/Dummett-style, while noncontextualized $x \sim y$ is equivalent to something of the form

$$| \, Size(y) - Size(x)| < n.$$

Then what we see is the usual pattern:

1. $Smoll(x) \ \& \ x \sim y \ \rightarrow \ Smoll(y)$.
2. $Smoll(x) \ \& \ (x \sim y)^A \ \rightarrow \ Smoll \ (y)$.

Version **1** is clearly invalid and supportive. To determine the status of version **2**, one has to assume that $I \in A$, which seems natural enough.[23] Consequently, version **2** becomes valid for perfectly familiar reasons: suppose $S(x)$ and $\neg S(y)$, then $x \prec I$, but not $y \prec I$, which implies

---

[23] For example, it is reasonable to assume that *if* the speaker is not an element of the comparison set that is active when processing of the premiss starts, *then* the clause $Smoll(x)$ adds the speaker to the comparison set.

$\neg(x \sim y)^A$.[24] This shows that a vague predicate need not be dependent on discourse context to give rise to the kinds of ambiguities that we have been concerned with in this paper, and which help us to understand the sorites paradox.

This example may be put aside on the grounds that *smoll* is context-dependent, since a person's *smoll*ness depends on the size of the speaker. I believe that this objection is not justified, since absolute judgments involving a vague predicate can always be re-analysed as involving an implicit comparison with other objects. Let me illustrate this. Among the properties that disqualify one for the privilege of performing military service in the Dutch army is the property of being small in the sense of being smaller than 1.65m. The way in which this is tested is by comparison to a pole that is exactly 1.65m tall. If you are visibly less tall than the pole you are small; if not, you are not small and have to find other ways to get yourself disqualified. Thus, one may define:

$$Small(x) \Leftrightarrow_{Def} x \prec p,$$

and this makes the definition of smallness equivalent to that of *smoll*, provided the '*I*' featuring in the definition of *smoll* happens to be exactly 1.65m. tall. If it is assumed that p is a member of the initial comparison set, then even in this non-contextdependent case, valid/nonsupportive versions of the inductive premiss can be obtained.[25] The situation may be summarized by saying that even seemingly non-contextdependent vague predicates must always involve a comparison, and that the thing or things with which objects are compared can also be relevant for the question of whether two objects are distinguishable.

How satisfactory is the solution of the sorites paradox that we have offered? I believe that it fulfils the *first* requirement that was formulated in the Introduction, namely of explaining the invalidity of the paradox, while at the same time accounting for its plausibility. Also, it is very much in the spirit of the *third* requirement, since its departures from classical logic are modest, as we have just seen. The only qualification that I can see has to do with the *second* requirement, of empirical adequacy. It might be argued that our proposal fails to reflect an intuition that has sometimes been put forward: that an object a that is smaller than a certain small object b is small *to a higher degree* than b. This purported intuition of gradualness has been exploited to

---

[24]Note that this situation is almost exactly analogous to the one in the illustration of section 4 in which the paradox breaks down because $x_{10} \sim I$.

[25]Needless to say, valid/nonsupportive versions of **P** are also obtained through notions of indistinguishability that guarantee equality of size, or through a notion of indistinguishability that uses a 'safety valve'.

solve the sorites paradox, but it seems that this can only be done at the cost of a highly nonclassical logic.[26] There is no technical obstacle to superimposing gradualness – in the form of multiple truth values – on a proposal of the kind that was advocated in this paper. But what I have tried to show is that such a drastic move is not required for the solution of the sorites paradox.

7 Appendix: A language definition, and some theorems

1. A formal language. A formal language + semantics in which the paradox can be formulated can be set up along the following lines. I have here opted for the interpretation of 'smallness of x with respect to A' that conforms with the Comparison Postulate. There are sufficiently many variables x, y, z, etc., and individual constants a, b, c, etc. as names for all the individuals in the domain. Atomic formulas are only of the forms $(S(x))^A$ or $(x \sim y)^A$. Compound formulas are constructed using negation, conjunction, implication and universal quantification as usual:

> If ϕ and ψ are formulas, then $\neg\phi$, $\phi\&\psi$, $\phi \to \psi$, and $\forall x(\phi)$ are formulas.

Truth for atomic formulas has to reckon with relativizations, but otherwise the definition is classical. Let a model M consist of a universe E, a comparison set $A \subseteq E$, and an Interpretation function I. Then truth with respect to M is defined as follows:

> $\|(S(a))^A\|$ is true w.r.t. M iff $|\{x \in A : x \succ I(a)\}| > |\{x \in A : I(a) \succ x\}|$, and false w.r.t. M otherwise.
> $\|(a \sim b)^A\|$ is true w.r.t. M iff $I(a) \sim I(b)$ & $\forall z \in A :$ $z \prec I(a) \leftrightarrow z \prec I(b)$ and $\forall z \in A : z \succ I(a) \leftrightarrow z \succ I(b)$.
> $\|\neg\phi\|$ is true w.r.t. M iff $\|\phi\|$ is false w.r.t. M. Otherwise it is false w.r.t. M.
> $\|\phi \to \psi\|$ is false w.r.t. M iff $\|\phi\|$ is true w.r.t. M and $\|\psi\|$ is false w.r.t. M. Otherwise it is true w.r.t. M.
> $\|\forall x(\phi)\|$ is true w.r.t. M iff $\phi[x/a]$ is true w.r.t M, for all individual constants a.

An abbreviation allows suppression of the empty set as a comparison set: $x \sim y =_{Def} (x \sim y)^\emptyset$, or equivalently $(x \sim y)^{\{x,y\}}$.

2. Theorem: (Non)transitivity of several notions of indistinguishability. Clearly, mere perceptual indistinguishability is

---

[26] A well-known example of a 'gradual' logic is fuzzy logic (Zadeh 1975). Gradualness can enter a logic without making it wildly nonclassical (e.g., Kamp 1975), but the resulting systems seem as badly disposed to solve the sorites paradox as classical logic.

nontransitive. The same holds for the Goodman/Dummett version of indistinguishability, in the version in which $(x \sim y)^A$ logically implies $x \sim y$. Note that Goodman/Dummett indistinguishability would have been transitive if it would have been defined without a separate clause to guarantee that $(x \sim y)^A$ logically implies $x \sim y$:

$$(x \prec y)^A \Leftrightarrow$$
$$\exists h \in A : (h \prec y \ \& \ \neg h \prec x) \lor (x \prec h \ \& \ \neg y \prec h).$$

Proof: Assume $(x \sim y)^A$ and $(y \sim z)^A$ and suppose $\neg(x \sim z)^A$. Then a contradiction is inferred as follows. We know there exists a help-element $h_{x,z}$ in A such that *either* $x \prec h_{x,z}$ and $z \sim h_{x,z}$, *or* $h_{x,z} \prec z$ and $h_{x,z} \sim x$. Assume the first. Now y may either be indistinguishable from $h_{x,z}$ (i.e., $y \sim h_{x,z}$), and then $h_{x,z}$ behaves differently towards y and x, so $\neg(x \sim y)^A$; or otherwise y is *not* indistinguishable from $h_{x,z}$ (i.e., $y \prec h_{x,z}$) but then $h_{x,z}$ behaves differently towards y and z, so $\neg(y \sim z)^A$. The second case (in which $h_{x,z} \prec z$ and $h_{x,z} \sim x$) is proven analogously. $\square$

A simple extension of the same proof also shows that the following weakening of transitivity is valid for the version of Goodman/Dummett indistinguishability that we have used:

$$\forall x, y, z, A \ (x, y, z \in A \rightarrow$$
$$((x \sim y)^A \ \& \ (y \sim z)^A \rightarrow (x \sim z)^A)).$$

3. No hierarchy of indistinguishability. For a given comparison set, one might, in principle, define an infinite series of notions $\sim_0$ (i.e., simple, perceptual indistinguishability, i.e., $\sim$), $\sim_1$ (i.e., Goodman/Dummett-style indistinguishability), $\sim_2$, etc. , using the following construction:

$$x \sim_0 y \Leftrightarrow_{Def} \neg(x \prec y) \ \& \ \neg(y \prec x),$$
$$x \sim_1 y \Leftrightarrow_{Def} x \sim_0 y \ \& \ \{z : z \sim_0 x\} = \{z : z \sim_0 y\},$$
$$x \sim_2 y \Leftrightarrow_{Def} x \sim_1 y \ \& \ \{z : z \sim_1 x\} = \{z : z \sim_1 y\},$$
$$\ldots,$$
$$x \sim_i y \Leftrightarrow_{Def} x \sim_{i-1} y \ \& \ \{z : z \sim_{i-1} x\} = \{z : z \sim_{i-1} y\}$$

Now obviously, $\sim_1$ differs from $\sim_0$, but for $n > 0 : \sim_n = \sim_1$. To prove this, it is sufficient to show that $\sim_2 = \sim_1$. But this fact follows from the following Lemma: (Notation: $aHbc$ means a is a help-element – in the sense of $\sim_1$ – that allows one to infer a difference between b and c.)

Lemma: $hHxz \& \neg hHyz \rightarrow hHxy$

Proof: One case that verifies the first premiss obtains when $x \prec h$ and $z \sim h$. The second premiss implies that $z \sim h$ $\Rightarrow y \sim h$, so it must be true that $y \sim h$. But if $y \sim h$ while $x \prec h$ then $hHxy$. The other cases verifying the first premiss are handled by a similar pattern of reasoning. $\square$

4. 'Almost-validity'. Here is a bidirectional version of the statement in the text, that a certain *in*valid version of the inductive premiss is valid under some additional restrictions on models.

Theorem: Given the Comparison Postulate, $\forall x, y : (S(x)^A$ & $x \sim y) \rightarrow S(y)^A$ is true for a given comparison set A iff $\forall x, y, A : (x \sim y \rightarrow (G[x]^A = G[y]^A$ & $K[x]^A = K[y]^A))$.

Proof: The right-to-left direction ($\Leftarrow$) is immediate. Conversely, ($\Rightarrow$) assume $\forall x, y, A : (S(x)^A$ & $x \sim y) \rightarrow S(y)^A$ is true in all admissible interpretations of S, while, for certain x, y, A such that $x \sim y$, $\exists z \in A$ such that $z \in G[x]^A - G[y]^A$. Then A' can be constructed such that $S(x)^{A'}, \neg S(y)^{A'}$. Example: $A' := \{z\}$, so $G[x]^A = \{z\}$ and $K[x]^{A'} = \emptyset$, while $G[y]^{A'} = K[y]^{A'} = \emptyset$. $\square$

5. No unlimited S-sequences.

Theorem: Assume there are no infinite comparison sets, while S satisfies constraints $S_1 - S_3$. Then there cannot exist unlimited S-sequences in any model.

Proof: Assume there exists an Unlimited S-sequence $S(x_0)^{A_0}$, $S(x_1)^{A_0 \cup \{x_0\}}, ..., S(x_i)^{A_0 \cup \{x_0, .., x_{i-1}\}}, ...$, where A_0 is finite. Then S_1 asserts that some other element y must fail to be small with respect to A_0: $\neg S(y)^{A_0}$. But S_2 says that adding an element to A_0 that is small with respect to A_0 does not change the status of y, so $\neg S(y)^{A \cup \{x_0\}}$. This procedure may be repeated any number of times, so one has $\neg S(y)^{A_0 \cup \{x_0, .., x_n\}}$. Consequently, elements that are even bigger than y will be not-small as well with respect to $A_0 \cup \{x_0, .., x_n\}$ (S_3), so Size(y) will be an upper limit of the kind that cannot exist if the sequence $S(x_0)^{A_0}$, $S(x_1)^{A_0 \cup \{x_0\}}, ..., S(x_i)^{A_0 \cup \{x_0, .., x_{i-1}\}}, ...$ is an Unlimited S-sequence. $\square$

6. Unlimited $\sim$-paths lead to unlimited S-sequences.

Theorem: Assume $S(x_0)^{A_0}$, for certain x_0. Then if a **P**-version $\langle P, MP \rangle$ logically implies the existence of unlimited $\sim$-paths in M, then it logically implies the existence of unlimited S-sequences in a model M.

Proof: Immediate. $\square$

7. A strenghening of S_2. As it stands, S_2 does not guarantee that the addition to a comparison set of individuals that are small with respect to that comparison set will have any effect on the norms for smallness. But plausible strengthenings are possible. For example, let $gk(A)$ denote the greatest element of A that is still small with respect to A: so $S(gk(A)^A$, but for all $y > gk(A)$, it holds that $\neg S(y)^A$. Then it is easy to prove the following theorem:

> *Theorem:* Suppose the Comparison Postulate holds, and assume A is finite. Assume further that for some x, $S(x)^A$. Then after adding finitely many elements $x_1, x_2, ..., x_n$ to A (the result of which is called A') all of which are visibly smaller than $gk(A)$, the result will be that now $\neg S(gk(A))^{A'}$.
>
> *Proof:* Immediate. □

8. The truth of $S_1 - S_3$ for various notions of smallness.

i. $S(x)^A \Leftrightarrow_{Def} |G[x]^A| > |K[x]^A|$. (**Comparison Postulate**)

Constraint S_1. If A is finite, then $G[x]^A$ is also finite and has a greatest element y. But then $\neg S(y)^A$, since $G[y]^A = \emptyset$. □

Constraint S_2. *Theorem:* Let $A' = A \cup \{x\}$. Assume that $S(x)^A$. Then $\|S^{A'}\| \subseteq \|S^A\|$. *Proof:* Assume $z \notin \|S^A\|$. Since $z \notin \|S^A\|$ but $x \in \|S^A\|$, we know that $x \notin G[z]^A$, so $G[z]^A = G[z]^{A'}$. But $A \subseteq A'$ implies $K[z]^A \subseteq K[z]^{A'}$.

$$K[z]^A \qquad \geq \qquad G[z]^A$$

$$\cap \qquad\qquad \|$$

$$K[z]^{A'} \qquad\qquad G[z]^{A'}$$

But $z \notin \|S^A\|$, so $K[z]^A \geq G[z]^A$, and therefore (see picture) $K[z]^{A'} \geq G[z]^{A'}$. It follows that $z \notin \|S^{A'}\|$. □

Constraint S_3. If $G[x]^A \leq K[x]^A$ and $x < y$ then $G[y]^A \leq K[y]^A$, since $G[y]^A \leq G[x]^A$ and $K[y]^A \geq K[x]^A$. □

ii. $S(x)^A \Leftrightarrow_{Def} |G[x]^A| > 1/2|A|$. (**Modified Comparison Postulate.**) Again, all three constraints are true. The proofs are exactly analogous to those for **i.** □

iii. $S(x)^A \Leftrightarrow_{Def}$ Size(x) is smaller than the average of the sizes of elements of A. (**Average Size postulate.**) As before, $S_1 - S_3$ are fulfilled. Let $Av(A)$ denote the average size of the elements of A. Then (S_1) there is an element x such that Size$(x) \geq Av(A)$. Consequently, $\exists x \neg S(x)^A$, which is even stronger than S_1. Further, (S_2) let $S(y)^A$, i.e., Size$(y) < Av(A)$. Then $Av(A \cup \{y\}) < Av(A)$, so $\|S^{A \cup \{y\}}\| \subseteq \|S^A\|$.

Finally, (S_3) if $\text{Size}(x) \geq Av(A)$, and $x < y$, then $\text{Size}(y) \geq Av(A)$, so $\neg S(y)^A$. □

9. A class of valid premisses.

Theorem: Assume indistinguishability is defined as by Goodman/Dummett, then the inductive premiss (in any version in which the three clauses are relativized to either A or $A \cup \{x\}$ or $A \cup \{y\}$ or $A \cup \{x, y\}$) is valid if and only if for all x and A, $S(x)^A \Leftrightarrow \Theta(G[x]^A, K[x]^A)$, for some relation Θ between sets.

Proof: ($\Rightarrow$) Suppose $S(x)^A \Leftrightarrow \Theta(G[x]^A, K[x]^A)$, and assume $S(x)^A$ and $(x \sim y)^A$, while $\neg S(y)^A$. Then either $G[x]^A \neq G[y]^A$ or $K[x]^A \neq K[y]^A$, but in either case, the assumption $(x \sim y)^A$ is contradicted. Conversely ($\Leftarrow$), suppose the inductive premiss is valid, while indistinguishability is defined following Goodman/Dummett. Suppose $S(x)$ is *not* co-extensional with a relation holding between $G[x]^A$ and $K[x]^A$. Therefore, there exist x, y and A such that $S(x)^A$ and $\neg S(y)^A$, even though $G[x]^A = G[y]^A$ and $K[x]^A = K[y]^A$. Consequently, $(x \sim y)^A$, which implies that the inductive premiss is not valid, contradicting the assumptions. □

References

Barwise, Jon. 1987. Noun Phrases, Generalized Quantifiers and Anaphora. In *Generalized Quantifiers*, ed. P.Gärdenfors. Also appeared as Report No. CSLI-86-52, CSLI, Stanford University 1986.

Van Benthem, Johan. 1983. *The Logic of Time*. Dordrecht, The Netherlands, Boston, USA, and London, UK: Reidel.

Buvač, Saša. 1993. Propositional Logic of Context. internal report, Stanford Univ., computer science Dept.

Carter, David. 1987. *Interpreting anaphors in natural language texts*. New York: Ellis Horwood, Wiley & Sons.

Condoravdi, Cleo, and Gawron, M. 1994. The context-dependency of implicit arguments. (This volume.)

Van Deemter, Kees. 1992. Towards a Generalization of Anaphora. *Journal of Semantics* 9: 27–51.

Van Deemter, Kees. 1994. The role of ambiguity in the sorites fallacy. *Proceedings of the Amsterdam Colloquium*, December 1993.

Dummett, Michael. 1975. Wang's Paradox. *Synthese* 30:301–324.

Goodman, Nelson. 1966. *The structure of appearance*. Dordrecht, The Netherlands and Boston, USA: Reidel.

Groenendijk, Jeroen, and Stokhof, M. 1991. Dynamic Predicate Logic. *Linguistics and Philosophy* 14: 39–100.

Heim, Irene. 1992. *The semantics of definite and indefinite Noun Phrases*. Ph.D.Thesis, University of Massachusetts, Amherst, Mass.

Hilbert, David. 1987. *Color and color perception*. Stanford: CSLI Lecture Notes No.9.

Kamp, Hans. 1975. Two theories about adjectives. In *Semantics for Natural Language*, ed. E.Keenan. Cambridge: Cambridge University Press.

Kamp, Hans. 1981a. A theory of truth and semantic interpretation. In *Formal Methods in the Study of Language*, ed. J.Groenendijk, T.Janssen, M.Stokhof. Mathematical Centre Tracts No.136.

Kamp, Hans. 1981b. The paradox of the heap. In *Aspects of Philosophical Logic*, ed. U.Mönnich.

Kaplan, David. 1979. On the logic of demonstratives. In *Contemporary Perspectives on the philosophy of language*, ed. P.A.French, T.E.Uehling Jr., and H.K.Wettstein, 401–412. Minneapolis: University of Minnesota Press.

Klein, Ewan. 1980. A semantics for positive and comparative adjectives. *Linguistics and Philosophy* 4.

Kwast, Karen. 1981. De sorites paradox. Masters thesis, University of Amsterdam.

McCarthy, John. 1987. Generality in Artificial Intelligence. *Communications of the ACM* 30:1030–1035. Also in *Turing Award Lectures, The First Twenty Years*, ACM Press 1987.

Montague, Richard. 1974. Pragmatics. In *Selected papers of Richard Montague*, ed. R.T.Thomason. New Haven and London: Yale University Press.

Partee, Barbara. 1989. Binding Implicit Variables in Quantified Contexts. *Papers of the Chicago Linguistic Society* 25: 342–365.

Suppes, Pattrick, and Zinnes, J.L. 1963. Basic Measurement Theory, In *Handbook of Mathematical Psychology*, ed. R.D.Luce et al., Vol. i:1–76. New York: Wiley.

Veltman, Frank. 1987. Logische Analyse, Course material for the University of Amsterdam, Dept. of Philosophy.

Westerståhl, Dag. 1985. Determiners and context sets. In *Generalized Quantifiers in Natural Language*, ed. J.van Benthem and A.ter Meulen. Dordrecht, The Netherlands: Foris, GRASS-4.

Zadeh, Lotfi. 1975. Fuzzy Logic and Approximate Reasoning. *Synthese* 30: 407–428.

4

Presuppositions and Information Updating

JAN VAN EIJCK

1 Introduction

Presupposition failures are errors occurring during the left-right processing of a computer program or natural language text. A general method for analysing such errors with dynamic logic is presented, based on the idea that sequential processing changes context dynamically and that this process of context change can be made the object of analysis in dynamic modal logic.

Updating a state of information with a statement bearing a presupposition can be viewed as a combination of two things: (i) checking whether the presupposition holds in the current context, and, provided this is the case, (ii) updating the current state of information with the informational content of the statement. In case the presupposition is not fulfilled in the current context, one might adjust the context to make it hold, and next do the further update with the informational content of the statement. These two actions are obviously *ordered*. First the context is adjusted to accommodate the presupposition, next the information state is updated with the assertion. Context and information state are used interchangeably here, and indeed, we can view the context which gets updated as a state of information.

We propose to view the study of presupposition failure from the general perspective of updating information structures (van Benthem 1991, de Rijke 1994, Jaspars to appear). In this perspective, providing information is a dynamic process involving a speaker and an audience, which gets the audience from an initial information state to a new more

I would like to thank Jan Jaspars for some illuminating conversations on the topic of information updating, and Makoto Kanazawa for his very helpful comments on two earlier versions of this paper.

Quantifiers, Deduction, and Context
Makoto Kanazawa, Christopher Piñón, and Henriëtte de Swart, editors
Copyright © 1996, CSLI Publications

informed state, or in case the new information is inconsistent with the current state, to the absurd information state.

It turns out that in many cases the information update relation is functional. In such cases presupposition failure can be catered for by switching to a partial update function. In case the presupposition of some update is not met in some information state, the update function yields 'undefined' for that update in that state. But information updating is not always functional. Communication mismatches do occur if the new information is too vague to yield a unique update in the current information state. In such cases the audience will have to indicate that the update cannot be processed without further ado. Note that this is different from the communication mismatch which occurs if the new information hinges upon a wrong assumption about the current context (the cases of presupposition failure).

The point that presupposition has to do with information updating has been made again and again in the literature. Stalnaker and Karttunen come to mind as early proponents of the view that presupposition projection should be accounted for in dynamic terms. See Stalnaker (1972, 1974), Karttunen (1973, 1974), Karttunen and Peters (1979), and Heim (1983). Modern versions of this approach appear in Beaver (1992), van Eijck (1994), Krahmer (1994), and Zeevat (1992).

Stalnaker proposes the following explanation of the rule (first stated in Karttunen 1973) that the presupposition of a conjunction $A\&B$ consists of the presupposition of A conjoined with the implication $\text{ass}_A \rightarrow \text{pres}_B$.

> The explanation goes like this: ... when a speaker says something of the form A *and* B, he may take it for granted that A ... after he has said it. The proposition that A will be added to the background of common assumptions before the speaker asserts that B. Now suppose that B expresses a proposition that would, for some reason, be inappropriate to assert except in a context where A, or something entailed by A, is presupposed. Even if A is *not* presupposed initially, one may still assert A *and* B since by the time one gets to saying that B, the context has shifted, and it is by then presupposed that A.
>
> From Stalnaker (1974), p. 211, also quoted in Heim (1992).

Modern versions of the dynamic approach to presupposition all attempt to formalize the *pragmatic* notion of presupposition, i.e., the notion of presupposition where a presupposition is a presupposition of the speaker about the context when he/she utters a sentence in that context, rather than a property of the sentence itself. On the other hand, inappropriateness of a sentence in a given context can be construed as lack of a definite truth value of that sentence in that context, and the presupposition of a sentence can always be viewed as a statement which holds in precisely those contexts where the sentence is appropriate.

We should distinguish, therefore between talking about what holds in given states of information on one hand and about updating information states on the other. It is precisely here that dynamic logic, which distinguishes between a level of static description and a level of procedural description, and which provides means of relating these two description levels, becomes an illuminating tool. The difference between the present analysis and other recent dynamic approaches to presupposition is the focus on the *link* between statics and dynamics, which relates semantic presupposition to pragmatic presupposition.

2 Information Structures

An information structure I is a pair $\langle S, \sqsubseteq \rangle$ with S a non-empty set of information states and $\sqsubseteq$ a pre-order (transitive and reflexive, but not necessarily antisymmetric) over S which is called the information order.

If L is a language for S, then an L-information model M is a triple $\langle S, \sqsubseteq, \sigma \rangle$, where $\sigma : L \to \mathcal{P}S$ is a specification function which interprets the language L in S. Classical languages can be specified by a single function σ, but for languages of partial logic one needs pairs σ^+, σ^- of such functions, and so on.

We consider the simplest case first, the case of propositional logic, where the states of information are sets of propositional valuations. Let a set of proposition letters P be given. Then the set of valuations is the set $\{0,1\}^P$. Call this set W. Members of W may be considered as '(epistemically) possible worlds'. An information state is a subset of W plus a member of W (the distinguished member plays the role of the 'actual world' or the 'current perspective of the knowing subject'); the set S of all information states is $\{\langle i, w \rangle \mid i \subseteq W, w \in W\}$. The information ordering $\sqsubseteq$ on S is given by:

$$\langle i, w \rangle \sqsubseteq \langle j, w' \rangle \text{ iff } i \supseteq j \text{ and } w = w'.$$

Note that this gives a partial order, not just a pre-order. Note also that we do not demand $w \in i$ if $\langle i, w \rangle$ is an information state. In particular, $\langle \emptyset, w \rangle$ is an information state, namely the absurd information state, viewed from perspective w. Without 'perspective worlds' it would be more awkward to modally characterize inconsistent information. With the use of a perspective world, we can simply say that $\Box \bot$ characterizes an inconsistent belief state.

Information states can be viewed as K45 models: sets of possible worlds with an 'almost' universal accessibility relation, i.e., transitive and almost reflexive and almost symmetric, for the 'perspective world' need not be accessible from itself. To be precise, K45 logic is complete for transitive and euclidean frames (a relation R is euclidean if xRy and xRz together imply that yRz),

K45 models are appropriate to talk about the kind of belief where one has complete information about one's own uncertainty (for more information about this, see Moore 1984). An appropriate 'local' language L to talk about information states is the language of propositional modal logic:

$$\alpha \quad ::= \quad \bot \mid p \mid \neg\alpha \mid (\alpha_1 \wedge \alpha_2) \mid \Diamond\alpha.$$

We employ the usual abbreviations for $\top, \wedge, \rightarrow, \leftrightarrow, \Box$. The specification function σ for the language is given by:

$$
\begin{aligned}
\sigma(\bot) &= \emptyset, \\
\sigma(p) &= \{\langle i, w\rangle \in S \mid w(p) = 1\}, \\
\sigma(\neg\alpha) &= S - \sigma(\alpha), \\
\sigma(\alpha_1 \wedge \alpha_2) &= \sigma(\alpha_1) \cap \sigma(\alpha_2) \\
\sigma(\Diamond\alpha) &= \{\langle i, w\rangle \in S \mid \exists w' \in i \text{ with } \langle i, w'\rangle \in \sigma(\alpha)\}.
\end{aligned}
$$

We say that $s \models \alpha$ iff $s \in \sigma(\alpha)$. Note that $\langle i, w\rangle \models \Box\bot$ iff $i = \emptyset$. An information state $\langle \emptyset, w\rangle$ is absurd or inconsistent: if we are in such a state, nothing at all is compatible with what we know or believe. Any boxed formula is true in an absurd information state; indeed, $\langle i, w\rangle \neq \langle \emptyset, w\rangle$ iff there is a formula $\alpha \in L$ with $\langle i, w\rangle \not\models \Box\alpha$.

An information state $\langle W, w\rangle$ is an information state of complete ignorance, a state of having no information at all. If P is infinite, there is no formula of L which characterizes W; for finite P there is. If $P = \{p_0, \ldots, p_n\}$, let C be the (finite) set of all conjunctions of form $\Diamond((\neg)p_0 \wedge \cdots \wedge (\neg)p_n)$. Then:

$$\langle i, w\rangle \models \bigwedge C \text{ iff } i = W.$$

3   Updating Propositional Information

Updating a state of information with a new piece of information can be viewed as moving up in the information order, toward some more informed state. Propositional updates are just tests (performed in the current perspective world). Epistemic updates can shift the information state: updating the information of one's audience with $\Box F$ will get the audience in a state where F is known, i.e., a state where $\Box F$ holds. Similarly, downdating with $\Box F$ will get the audience in a state where F is not known anymore, i.e, in a state where $\Diamond\neg F$ holds.

In the more general perspective on information structures, neither updates or downdates need to be minimal (cf. van Benthem 1991 and de Rijke 1994), but for present purposes this restriction is useful. Minimal updates are given by:

$$[\![\alpha^u]\!] = \{\langle s, s'\rangle \in S \times S \mid s \sqsubseteq s', s' \in \sigma(\alpha),$$
$$\forall s'' \in S : (s \sqsubseteq s'' \sqsubseteq s' \ \& \ s'' \in \sigma(\alpha)) \Rightarrow s' \sqsubseteq s''\}$$

We will use F, G, F_1, ... as metavariables for purely propositional formulas of L. As it turns out, the minimal update relation for L is functional for updates of the form $\Box F$ or $\Diamond F$.

We may assume that knowing subjects, unless they are all-knowing, cannot distinguish the actual world from any other of their epistemic alternatives, so updates with purely propositional F are a bit silly: they can succeed only if they do not change states, and their success depends on what is true in the actual world.

An update with $p \vee q$ in state $\langle i, w\rangle$ checks if $i, w \models p \vee q$, and succeeds if this is the case, fails otherwise. In other words, this update succeeds iff $w(p) = 1$ or $w(q) = 1$.

If I utter $p \vee q$, this should be understood as: 'I believe or know that $p \vee q$, and I want you to accept that information about the world, too. So, this update should be understood as having an implicit $\Box$ in front.

An update with $\Box(p \vee q)$ in state $\langle i, w\rangle$ causes a shift to a state $\langle j, w\rangle$ where $j = \{w' \in i \mid \langle i, w'\rangle \models p \vee q\}$.

An update with $\Diamond(p \vee q)$ in state $\langle i, w\rangle$ does not change state in case $\langle i, w\rangle \models \Diamond(p \vee q)$, and otherwise fails.

Updates with $\Diamond F$ formulas are 'consistency tests': they check whether the information F is consistent with the current information state. Let $\|\alpha\|_i$ be $\{w \in W \mid \langle i, w\rangle \in \sigma(\alpha)\}$. The following proposition holds:

Proposition 1

1. $[\![F^u]\!] = \{\langle s, s\rangle \in S \times S \mid s \models F\}$.
2. $[\![\Box F^u]\!] = \{\langle\langle i, w\rangle, \langle i \cap \|F\|_i, w\rangle\rangle \mid \langle i, w\rangle \in S\}$.
3. $[\![\Diamond F^u]\!] = \{\langle s, s\rangle \in S \times S \mid s \models \Diamond F\}$.

An update with $\Box p \vee \Box q$ in state $\langle i, w\rangle$ is not functional in case i contains a w_1 with $w_1(p) = 1, w_1(q) = 0$ and a w_2 with $w_2(p) = 0, w_2(q) = 1$. In this case there are two possible outcome states: $s_1 = \langle\{w' \in i \mid w'(p) = 1\}, w\rangle$ and $s_2 = \langle\{w' \in i \mid w'(q) = 1\}, w\rangle$. Take for example the case where $s = \langle\{p\bar{q}, \bar{p}q\}, w\rangle$ (where $\bar{p}$ indicates that p is false). Both

$$\langle\{p\bar{q}, \bar{p}q\}, w\rangle \mapsto \langle\{p\bar{q}\}, w\rangle$$

and

$$\langle\{p\bar{q}, \bar{p}q\}, w\rangle \mapsto \langle\{\bar{p}q\}, w\rangle$$

are minimal updates.

Modal updates are discussed in Veltman (1991), but with the constraint that only modal updates of the forms $\Diamond\top \to \Diamond F$ and $\Box F$ are allowed. Our $\Diamond\top \to \Diamond F$ corresponds to Veltman's *might F*, and our $\Box F$ to his F, so the $\Box$ is left implicit in his notation. Updates of the form $\Diamond\top \to \Diamond F$ are total functions, while updates of the form $\Diamond F$ may be partial. In fact, an update with $\Diamond\top \to \Diamond F$ will effect a transition to an inconsistent state if $\Diamond F$ does not hold in the current state. Any update of the form $\Box F \wedge \bigvee \bigwedge \Diamond F_i$ (where F and all the F_i are purely propositional), is functional. An important result about K45, by the way, is that every formula has an equivalent formula of the form $\bigvee (F \wedge \Box G \wedge \bigwedge \Diamond H_i)$ (where F, G and the H_i are all purely propositional). This follows from the completeness of K45 with respect to finite 'balloon' frames, i.e., frames where the accessibility relation is transitive and euclidean; these frames have the shape of a balloon of mutually accessible worlds all accessible from a single perspective world. See e.g. Chellas (1980) for more information. Disjunction over $\Box$ is the feature that 'threatens' functionality.

The functional K45 formulas are 'honest' formulas in the sense of Halpern and Moses (1985). A formula is honest if one can honestly claim that one *only* knows that formula. This is equivalent to saying that a minimal update of the state of complete ignorance with that formula is functional. Claiming that you only know $\Box p \vee \Box q$ is a cheat, because in order for that to be a true statement you either have to be in a state where all the accessible worlds are p worlds, and in that case you also know p, or you have to be in a state where all the accessible worlds are q worlds, and in that case you also know q.

A K45 formula α is persistent if $\langle i, w \rangle \models \alpha$ and $\langle i, w \rangle \sqsubseteq \langle j, w \rangle$ together imply that $\langle j, w \rangle \models \alpha$. Formulas of the form $\Box F$ are persistent, formulas of the form $\Diamond F$, with F a consistent propositional formula, are not. Conjunctions and disjunctions of persistent formulas are persistent. Every persistent formula is K45 equivalent to a disjunction of conjunctions of formulas of the forms F and $\Box F$ (F purely propositional).

$\Diamond p$ is true at $\langle W, w \rangle$, but false at any information state $\langle i, w \rangle$ without p worlds.

The persistent and functional K45 formulas are precisely the formulas of the form $F_1 \wedge \Box F_2$, F_1 and F_2 purely propositional (up to K45 equivalence).

Conversely, we can look at minimal downdates to move back to a state where α does not hold anymore. Minimal downdates are given by:

$$[\![\alpha^d]\!] \;=\; \{\langle s, s' \rangle \in S \times S \mid s' \sqsubseteq s,\, s' \notin \sigma(\alpha),$$
$$\forall s'' \in S : (s' \sqsubseteq s'' \sqsubseteq s \;\&\; s'' \notin \sigma(\alpha)) \Rightarrow s'' \sqsubseteq s'\}.$$

The minimal downdate relation is not functional for formulas of the form $\Box F$, and it is a test for formulas of the form $\Diamond F$.

$[\![(\Box p)^d]\!]$ will relate a state $\langle i, w \rangle$ satisfying $\langle i, w \rangle \models \Box p$ to any state $\langle j, w \rangle$ where j is of the form $i \cup \{w'\}$, with $w'(p) = 0$, and a state $\langle i, w \rangle$ not satisfying $\langle i, w, \rangle \models \Box p$ to itself.

Downdates with formulas of the form $\Diamond F$ can only succeed in case $\Diamond F$ is inconsistent with the current information state.

$$[\![(\Diamond p)^d]\!] = \{\langle s, s \rangle \in S \times S \mid s \models \neg \Diamond p\}.$$

In general we have:

Proposition 2

1. $[\![F^d]\!] = \{\langle s, s \rangle \in S \times S \mid s \models \neg F\}.$
2. $[\![\Box F^d]\!] = \{\langle s, s \rangle \in S \times S \mid s \models \neg \Box F\} \cup \{\langle \langle i, w \rangle, \langle i \cup \{w'\}, w \rangle \rangle \mid \langle i, w \rangle \in S, \langle i, w \rangle \models \Box F, w' \in W - \|F\|_i\}.$
3. $[\![\Diamond F^d]\!] = \{\langle s, s \rangle \in S \times S \mid s \models \neg \Diamond F\}.$

Over the local language L we now layer a global language L_1, which is the language of the information structure S. L has a procedural and a propositional level; the procedures are (minimal) updating, (minimal) downdating, testing, plus sequential compositions of those, the propositions of L_1 are built from the formulas of L (which act as atoms of L_1) using boolean combination and procedure projections.

$$\alpha \quad ::= \quad \bot \mid p \mid \neg\alpha \mid (\alpha_1 \wedge \alpha_2) \mid \Diamond\alpha$$
$$\pi \quad ::= \quad \alpha^u \mid \alpha^d \mid \varphi? \mid (\pi_1; \pi_2)$$
$$\varphi \quad ::= \quad \alpha \mid \neg\varphi \mid (\varphi_1 \wedge \varphi_2) \mid \Diamond\varphi \mid \text{dom}\,(\pi) \mid \text{ran}\,(\pi) \mid \text{fix}\,(\pi).$$

The interpretation of L_1 consists of two parts: a relational interpretation for the procedures and a truth definition for the formulas. The interpretation for the procedures and formulas uses mutual recursion.

$$\begin{aligned}
[\![\alpha^u]\!] &= \quad \text{as given above.} \\
[\![\alpha^d]\!] &= \quad \text{as given above.} \\
[\![\varphi?]\!] &= \quad \{\langle s, s \rangle \in S \times S \mid S, s \models \varphi\}. \\
[\![\pi_1; \pi_2]\!] &= \quad [\![\pi_1]\!] \circ [\![\pi_2]\!].
\end{aligned}$$

In the final clause, $\circ$ denotes relational composition.

$$\begin{aligned}
S, s \models \alpha & \quad \text{iff} \quad s \models \alpha \\
S, s \models \neg\varphi & \quad \text{iff} \quad S, s \not\models \varphi \\
S, s \models (\varphi_1 \wedge \varphi_2) & \quad \text{iff} \quad S, s \models \varphi_1 \text{ and } S, s \models \varphi_2 \\
S, s \models \Diamond\varphi & \quad \text{iff} \quad s = \langle i, w \rangle \text{ and} \\
& \qquad \text{there is some } w' \in i \text{ with } S, \langle i, w' \rangle \models \varphi \\
S, s \models \text{dom}\,(\pi) & \quad \text{iff} \quad \exists s' \in S : s[\![\pi]\!]s' \\
S, s \models \text{ran}\,(\pi) & \quad \text{iff} \quad \exists s' \in S : s'[\![\pi]\!]s \\
S, s \models \text{fix}\,(\pi) & \quad \text{iff} \quad s[\![\pi]\!]s.
\end{aligned}$$

This system is an extension of the system of update logic presented in Veltman (1991) with tests, downdates and a more liberal regime concerning modal updates. Alternatively, it can be viewed as a fragment of a structured version of the Dynamic Modal Logic (DML) in van Benthem (1991), with structured states (in this case: K45 models) instead of unstructured propositional valuations, but with just a subset of the procedural repertoire.

We can define the perhaps more familiar dynamic logic style procedure modalities $\langle \pi \rangle$ and $[\pi]$ in terms of the projection operators and tests as follows:

$$\langle \pi \rangle \varphi \stackrel{\text{def}}{=} \text{dom}\,(\pi; \varphi?),$$

$$[\pi]\varphi \stackrel{\text{def}}{=} \neg\text{dom}\,(\pi; (\neg\varphi)?).$$

Note that it follows from these definitions that:

- $S, s \models \langle \pi \rangle \varphi$ iff $\exists s' \in S : s[\![\pi]\!]s'$ and $S, s' \models \varphi$.
- $S, s \models [\pi]\varphi$ iff $\forall s' \in S : s[\![\pi]\!]s'$ implies $S, s' \models \varphi$.

Note the following important differences:

$$
\begin{array}{lll}
S, s \models \bot & & \text{never.} \\
S, s \models \top & & \text{always.} \\
S, s \models \Box\bot & \text{iff} & s = \langle \emptyset, w \rangle. \\
S, s \models \Diamond\top & \text{iff} & s \neq \langle \emptyset, w \rangle.
\end{array}
$$

We see from the above that $[\pi]\bot$ expresses that no π transition is possible, while $[\pi]\Box\bot$ expresses that the only possible π transition will get one to the absurd information state, or, in other words, that the transition π yields inconsistency.

$\langle \pi \rangle \top$ expresses that a π transition is possible; $\langle \pi \rangle \Diamond \top$ expresses that a *consistent* π transition is possible.

The notion of validity for L_1 is as follows:

- $\models \varphi$ iff for all $s \in S : S, s \models \varphi$.

Because we have used the full space $\mathcal{P}W$ to define the state set S, there is no need to mention S as a parameter in the validity notion: once the set of proposition letters is fixed, the set of information states is fixed. This changes when we allow S to be a proper subset of

$$\{\langle i, w \rangle \mid i \in \mathcal{P}W, w \in W\},$$

subject to certain conditions. Nothing in the notion of 'information structure' prevents us from doing this, as long as we make sure that the information ordering $\sqsubseteq$ on S remains a pre-order. We will not explore this possibility here, however.

4 Some Example Validities

We will not present a full axiomatisation of the logic of propositional up- and downdating, but merely give some of the valid principles that we can use to reason about information transitions. Next to the obvious axioms and rules of inference of propositional logic, of normal modal logic for $\Diamond$, we need the following:

P 4.1 $\Box\varphi \to \Box\Box\varphi$.

P 4.2 $\Diamond\varphi \to \Box\Diamond\varphi$.

Principles 4.1 and 4.2 are the principles of positive and negative introspection of $K45$ for $\Diamond$.

P 4.3 dom $(\pi) \leftrightarrow$ dom $(\pi; \top?)$.

P 4.4 ran $(\pi) \leftrightarrow$ ran $(\top?; \pi)$.

These are to make sure that we can always assume dom arguments to be of the general form $\pi; \varphi?$, and ran arguments to be of the general form $\varphi?; \pi$.

P 4.5 $[\pi](\varphi_1 \to \varphi_2) \to ([\pi]\varphi_1 \to [\pi]\varphi_2)$.

This is the K schema for π, which expresses that for every information transition procedure π, the operator $[\pi]$ is a normal modal operator.

P 4.6 $\langle \alpha^u \rangle \top \leftrightarrow \langle \alpha^u \rangle \alpha$.

This expresses that after updating with α, α will hold. The principle does not entail that updates have the property of *right seriality* (for every s there is a t with $s[\![\pi]\!]t$), for we have seen that this need not be the case for α of the forms F or $\Diamond F$.

P 4.7 $[F^u]\varphi \leftrightarrow (F \to \varphi)$.

P 4.8 $[\Diamond F^u]\varphi \leftrightarrow (\Diamond F \to \varphi)$.

These express that F and $\Diamond F$ updates are tests.

P 4.9 $\langle \Box F^u \rangle \Box G \leftrightarrow [\Box F^u]\Box G \leftrightarrow \Box(F \to G)$.

The soundness of Principle 4.9 follows from the functionality of $\Box F$ updates, plus the next proposition.

Proposition 3 $s \models [\Box F^u]\Box G$ *iff* $s \models \Box(F \to G)$.

Proof. $\langle i, w \rangle \models [\Box F^u]\Box G$
iff $\langle i \cap \|F\|_i, w \rangle \models \Box G$
iff $\langle i \cap \{w' \mid w' \models F\}, w \rangle \models \Box G$
iff $\forall w' \in i$: if $w' \models F$ then $w' \models G$
iff $\langle i, w \rangle \models \Box(F \to G)$.   ∎

P 4.10 $\langle \Box F^u \rangle \Diamond G \leftrightarrow [\Box F^u]\Diamond G \leftrightarrow \Diamond(F \wedge G)$.

The soundness of Principle 4.10 follows from the functionality of $\Box F$ updates, plus the next proposition.

Proposition 4 $s \models [\Box F^u]\Diamond G$ *iff* $s \models \Diamond(F \wedge G)$.

Proof. $\langle i, w \rangle \models [\Box F^u]\Diamond G$
iff $\langle i \cap \|F\|_i, w \rangle \models \Diamond G$
iff $\langle i \cap \{w' \mid w' \models F\}, w \rangle \models \Diamond G$
iff $\exists w' \in i$: $w' \models F$ and $w' \models G$
iff $\langle i, w \rangle \models \Diamond(F \wedge G)$. ∎

P 4.11 $[F^d]\varphi \leftrightarrow (F \vee \varphi)$.

P 4.12 $[\Diamond F^d]\varphi \leftrightarrow (\Diamond F \vee \varphi)$.

These express that F and $\Diamond F$ downdates are tests.

The following principle is a rule rather than an axiom schema.

P 4.13 $\dfrac{(\bigwedge C \to \Diamond(F \wedge \neg G))}{(\text{ran}\,(\Diamond F?;\Box G^u) \leftrightarrow \Box G)}$.

Here $\bigwedge C$ is the conjunction of all formulas of the form $\Diamond((\neg)p_0 \wedge \cdots \wedge (\neg)p_n)$, where $p_0, \ldots, p_n$ are the proposition letters occurring in F, G.

P 4.14 $\text{ran}\,(\varphi?; F^u) \leftrightarrow (\varphi \wedge F)$.

P 4.15 $\text{ran}\,(\varphi?; \Diamond F^u) \leftrightarrow (\varphi \wedge \Diamond F)$.

P 4.16 $\text{ran}\,(\Box F?; \Box G^u) \leftrightarrow \Box(F \wedge G)$.

After updating a context with a persistent update, the persistent preconditions will hold in the new context.

P 4.17 $\text{ran}\,(\varphi?; F^d) \leftrightarrow (\varphi \wedge \neg F)$.

P 4.18 $\text{ran}\,(\varphi?; \Diamond F^d) \leftrightarrow (\varphi \wedge \neg\Diamond F)$.

P 4.19 $\text{ran}\,((\Diamond F_1 \wedge \cdots \wedge \Diamond F_n)?; \Box G^d) \leftrightarrow (\Diamond F_1 \wedge \cdots \wedge \Diamond F_n \wedge \Diamond\neg G)$.

The counterparts to the previous three for downdates.

P 4.20 $\langle \alpha^d \rangle \top \leftrightarrow \langle \alpha^d \rangle \neg \alpha$.

This expresses that a downdate can only succeed if after downdating with α, α does not hold anymore. Note that downdating will be impossible if the downdate is a logical validity (the present set-up is unsuitable for modelling 'unlearning' of logical truths).

P 4.21 $\text{fix}\,(\alpha^u) \leftrightarrow \alpha$.

This expresses that updating with α doesn't change the context precisely when α already holds.

P 4.22 $\text{fix}\,(\alpha^d) \leftrightarrow \neg \alpha$.

This expresses that downdates with information that is already known to be false have no effect.

P 4.23 $\text{fix}\,(\pi; \varphi?) \leftrightarrow \text{fix}\,(\varphi?; \pi) \leftrightarrow (\text{fix}\,(\pi) \wedge \varphi)$.

This is an obvious statement about fixpoints.

P 4.24 $(\text{fix}(\pi) \wedge \varphi) \rightarrow (\text{dom}(\pi; \varphi?) \wedge \text{ran}(\varphi?; \pi))$.

This relates fixpoint to domain and range.

P 4.25 $(\text{fix}(\pi_1) \wedge \text{fix}(\pi_2)) \rightarrow \text{fix}(\pi_1; \pi_2)$.

If s is a fixpoint for π_1 and π_2, then s is a fixpoint for $\pi_1; \pi_2$ (but note that this cannot be strengthened to an equivalence).

P 4.26 $\text{dom}(\pi_1; \pi_2; \varphi?) \leftrightarrow \text{dom}(\pi_1; \text{dom}(\pi_2; \varphi?)?)$.

This is the usual principle for sequential composition. Stated in terms of $\langle \pi \rangle$ it can also be expressed as $\langle \pi_1; \pi_2 \rangle \varphi \leftrightarrow \langle \pi_1 \rangle \langle \pi_2 \rangle \varphi$ (familiar from Pratt-style propositional dynamic logic).

P 4.27 $\text{ran}(\varphi?; \pi_1; \pi_2) \leftrightarrow \text{ran}(\text{ran}(\varphi?; \pi_1)?; \pi_2)$.

This is the counterpart to the previous principle. Finally, here are three axioms about testing.

P 4.28 $\text{dom}(\varphi_1?; \varphi_2?) \leftrightarrow (\varphi_1 \wedge \varphi_2)$.

P 4.29 $\text{ran}(\varphi_1?; \varphi_2?) \leftrightarrow (\varphi_1 \wedge \varphi_2)$.

P 4.30 $\text{fix}(\varphi?) \leftrightarrow \varphi$.

It should be noted that some of these principles can be simplified if we extend the relational repertoire of the language. For example, if we admit procedure intersection then we can define fixpoints by means of $\text{fix}(\pi) \leftrightarrow \text{dom}(\pi \cap \top?)$. Of course, the trade-off is that now extra principles for intersection have to be added.

5   Validity and Consequence

While there is an obvious static validity notion for the logic of propositional information transitions (see above), there are several candidates for the notion of 'dynamic valitity' (van Benthem 1991, Veltman 1991, van Eijck and de Vries 1995), which are all easily expressible in the present format.

An information transition π is always accepted if for every information state s, $\langle s, s \rangle \in [\![\pi]\!]$. This is the case iff for every information state s, $s \models \text{fix}(\pi)$.

An information transition π is always acceptable if for every information state $s \neq \langle \emptyset, w \rangle$, there is some $s' \neq \langle \emptyset, w \rangle$ with $\langle s, s' \rangle \in [\![\pi]\!]$. This is the case iff for every information state s, $s \models \Diamond \top \rightarrow \langle \pi \rangle \Diamond \top$.

Similarly, while it is clear what the 'static' notion of logical consequence should be (namely, $\Gamma \models \Delta$ iff for every $s \in S$ with $s \models \bigwedge \Gamma$ it holds that $s \models \bigvee \Delta$), there are several candidates for 'dynamic consequence' in this framework (van Benthem 1991; see Kanazawa 1994 for further analysis):

- $\pi_1 \models_1 \pi_2$ iff for all $s \in S$, $s[\![\pi_1]\!]s$ implies $s[\![\pi_2]\!]s$.

- $\pi_1 \models_2 \pi_2$ iff for all $s, s' \in S$ with $s[\![\pi_1]\!]s'$ and $s' \neq \langle \emptyset, w \rangle$ there is an $s'' \neq \langle \emptyset, w \rangle$ with $s'[\![\pi_2]\!]s''$.
- $\pi_1 \models_3 \pi_2$ iff for all $s, s' \in S$, $s[\![\pi_1]\!]s'$ implies $s'[\![\pi_2]\!]s'$.

These are readily expressed in terms of static validity, for we have:

- $\pi_1 \models_1 \pi_2$ iff $\models \text{fix}(\pi_1) \to \text{fix}(\pi_2)$.

And for the second one:

- $\pi_1 \models_2 \pi_2$ iff $\models [\pi_1](\Diamond\top \to \langle \pi_2 \rangle \Diamond\top)$.

And the third one:

- $\pi_1 \models_3 \pi_2$ iff $\models \text{ran}(\pi_1) \to \text{fix}(\pi_2)$.

6 Information Conveyed by an Update

One way of 'measuring' the information conveyed by an information transition in a state satisfying φ is by means of ran $(\varphi?; \pi)$.

The information conveyed by the update $(\Diamond p)^u; (\Box \neg p)^u$ (one of Veltman's key examples) in the state of complete ignorance is calculated as follows:

$$\text{ran}(\top?; (\Diamond p)^u; (\Box\neg p)^u) \quad \leftrightarrow \quad \text{ran}(\text{ran}(\top?; (\Diamond p)^u)?; (\Box\neg p)^u)$$
$$\leftrightarrow \quad \text{ran}(\Diamond p?; \Box\neg p^u)$$
$$\leftrightarrow \quad \Box\neg p.$$

The second step uses Principle 4.15, the third Principle 4.13.

The information conveyed by the update $(\Box\neg p)^u; (\Diamond p)^u$ in the state of complete ignorance is calculated as follows:

$$\text{ran}(\top?; (\Box\neg p)^u; (\Diamond p)^u) \quad \leftrightarrow \quad \text{ran}(\text{ran}(\top?; (\Box\neg p)^u)?; (\Diamond p)^u)$$
$$\leftrightarrow \quad \text{ran}(\text{ran}(\Box\top?; (\Box\neg p)^u)?; (\Diamond p)^u)$$
$$\leftrightarrow \quad \text{ran}((\Box\neg p)?; (\Diamond p)^u)$$
$$\leftrightarrow \quad \Box\neg p \wedge \Diamond p.$$

This is K45 equivalent to $\bot$, which shows that this update will never succeed. (Note the use of Principle 4.16 in the third step.)

In fact, since updating with Veltman's *might p* corresponds to updating with $\Diamond\top \to \Diamond p$, the 'rational reconstruction' of Veltman's example is slightly different:

The information conveyed by the update

$$(\Box\neg p)^u; (\Diamond\top \to \Diamond p)^u$$

in the state of complete ignorance is calculated as follows:

$$\text{ran}(\top?; (\Box\neg p)^u; (\Diamond\top \to \Diamond p)^u)$$
$$\leftrightarrow \text{ran}(\text{ran}(\top?; (\Box\neg p)^u)?; (\Diamond\top \to \Diamond p)^u)$$
$$\leftrightarrow \text{ran}(\text{ran}(\Box\top?; (\Box\neg p)^u)?; (\Diamond\top \to \Diamond p)^u)$$
$$\leftrightarrow \text{ran}((\Box\neg p)?; (\Diamond\top \to \Diamond p)^u)$$
$$\leftrightarrow \Box\neg p \wedge (\Diamond\top \to \Diamond p).$$

This is K45 equivalent to $\Box\bot$, which shows that this update will get the audience in an inconsistent state of information.

To see that the final step in the calculation is correct, note that the update procedure $(\Diamond\top \to \Diamond p)^u$ is equivalent to the procedure $(\Diamond\top?; \Diamond p^u) \cup \Box\bot^u$, where $\cup$ denotes choice between procedures. An obvious principle governing choice is:

$$\text{ran}\,(\pi_1; (\pi_2 \cup \pi_3)) \leftrightarrow (\text{ran}\,(\pi_1; \pi_2) \vee \text{ran}\,(\pi_1; \pi_3)).$$

Using this and the other principles, the final step can easily be validated.

In general, a transition π is consistent in state s iff $\langle\pi\rangle\Diamond\top?$ holds in s. This expresses that a transition from s via π is possible which does end up in a consistent state of information.

It is tempting, especially in the light of the dynamic consequence notion $\models_1$, to equate the information conveyed by an information transition π with fix (π). But note that the combination $\Diamond p^u; \Box\neg p^u$ does not have a fixpoint, while, as we have just seen, updating with that information starting from complete ignorance yields a *consistent* information state. Is there perhaps something funny about updates of the form $\Diamond F^u$?

From the present perspective, such updates are indeed strange. Recall that our intention is to model the knowledge of an *audience* addressed by a single speaker. If one assumes that $\Diamond F^u$ corresponds to an assertion by the speaker, then the update would have to correspond to an assertion about the state of knowledge of the audience: the speaker states that F is consistent with what the audience knows already. This assertion would correspond to something like 'I take it that you know that F is possible'. But this is not an assertion in the sense of 'statement influencing the state of knowledge of the audience'.

Compare this with the 'updates' of the form 'p may be the case', or 'maybe p' in Veltman (1991). Veltman renders the assertion 'maybe p' as an update with $\Diamond\top \to \Diamond p$. The big difference between Veltman's information states and ours is that Veltman's information states model the knowledge of a single agent reporting on how he or she processes incoming information, while ours model the knowledge of the *audience* addressed by a single speaker.

In our set-up, updates of the form $\Diamond\top \to \Diamond p$ are not consistency checks of one's own knowledge, as they are for Veltman, but statements about the knowledge of the audience. Since we cannot in general assume that a speaker has complete knowledge of what his or her audience believes, such statements are rather pointless. On the other hand, checking the knowledge of the audience by means of a test $\varphi?$ may still make eminent sense, as we will see in the next section.

In the present set-up, an assertion of the form 'maybe p' should be construed as an invitation to the audience to *reconsider* the truth of p, i.e., such an assertion is a *downdate*, and it has the form $(\neg \Diamond p)^d$, or equivalently $(\Box \neg p)^d$.

If one imposes the constraint that information transitions always be compositions of basic units of the forms $\Box F^u$ and $\Box F^d$, possibly interspersed with tests, then information content can always be described in terms of fixpoints. A consistent fixpoint for $\Diamond p^u; \Box \neg p^u$ does not exist, but for $\Box \neg p^d; \Box \neg p^u$ it does; indeed, any state where $\Box \neg p$ holds is such a fixpoint.

7 Expressing Presuppositions

We have seen that realistic information transitions in our set-up have the forms $(\Box F)^u$ or $(\Box F)^d$. In case such an information transition has a presupposition, we may assume that this has the form of a test to see whether something is known in the current context, i.e., a test of the form $\Box \varphi$? Not only updates may have presuppositions, witness (1).

(1)   Maybe the king of France is eating frog legs.

If I assert (1) then I presuppose that the king of France exists, and I invite my audience to revise their belief that his majesty is doing something other than eating frog legs. (We have seen that 'maybe statements' turn up as downdates of the form $\Box F^d$ in the present framework.) So this is a downdate with presupposition.

The framework presented above has all the machinery in place to express presuppositions. We can express the requirement that updating with α has presupposition φ in state s by means of the complex update $\varphi?; \alpha^u$.

$\Box p?; \Box q^u$ succeeds in state s iff $s \models \Box p$, and effects a transition to a state s' with $s \sqsubseteq s'$ and $s' \models \Box q$. (And of course also $s' \models \Box p$, for $\Box p$ is a persistent formula.)

The semantic clause for $[\![\varphi?; \alpha^u]\!]$ is given by:

$$
\begin{aligned}
[\![\varphi?; \alpha^u]\!] &= [\![\varphi?]\!] \circ [\![\alpha^u]\!] \\
&= \{\langle s, s' \rangle \in [\![\alpha^u]\!] \mid s \models \varphi\} \\
&= [\![\alpha^u]\!] - \{\langle s, s' \rangle \in S \times S \mid s \not\models \varphi\}.
\end{aligned}
$$

As the relational interpretation demonstrates, $\varphi?; \alpha^u$ is interpreted as an update with α under the presupposition that φ holds in the current context. Similarly, $\varphi?; \alpha^d$ is interpreted as a downdate with α under the presupposition that φ holds in the current context.

Calculating the presupposition of an information transition π consists in finding a specification of the information states s for which there is an s' with $\langle s, s' \rangle \in [\![\pi]\!]$. In these cases we say that the transition π

does not abort. Conversely, the presupposition failure conditions of an information transition π consist of a specification of the information states s for which there is no s' with $\langle s, s' \rangle \in [\![\pi]\!]$. We say in these cases that transition π *aborts* in state s.

To calculate the presupposition of an update α^u, we have to check the conditions on states s under which the relation $[\![\alpha^u]\!]$ does have a successor for s. These are given by the following schemata, which are derivable from the principles in the previous section:

T 7.1 $\langle \varphi?; \alpha^u \rangle \top \leftrightarrow \varphi \wedge \langle \alpha^u \rangle \top$.

This expresses that a simplex update with presupposition can be performed if and only if the presupposition holds in the current information state and the update without presupposition is possible in the current context.

T 7.2 $\langle \varphi?; \alpha^d \rangle \top \leftrightarrow \varphi \wedge \langle \alpha^d \rangle \top$.

This expresses that a downdate under presupposition is possible iff the presupposition holds in the current information state and the downdate without presupposition is possible in that state. Note that the presupposition of an information transition is nothing but the weakest preconditions for success of that transition, in the well known computer science sense.

For a concrete example, assume that the lexical presupposition of being a bachelor consists of being male plus being adult. We do not yet look inside the basic propositions built from these predicates, so we merely say that example (2) presupposes the conjunction of (3) and (4), and asserts (5).

(2)   Jan is a bachelor.

(3)   Jan is male.

(4)   Jan is adult.

(5)   Jan is unmarried.

Basic propositions that do not themselves have presuppositions can be represented using basic proposition letters. Let us use p for (3), q for (4), and $\neg r$ for (5).

The update for *Jan is a bachelor* does have the presuppositions *Jan is male* and *Jan is adult,* and (after the update with these presuppositions) it makes the assertion *Jan is unmarried,* so it can be represented as $\Box(p \wedge q)?; \Box \neg r^u$. Let P be the set $\{p, q, r\}$. Information states for this fragment are built from valuations in $\{p, q, r\} \rightarrow \{0, 1\}$.

(6)   Jan is male. Jan is a bachelor.

The meaning of the update with the sequence (6) is given by $[\![\Box p^u; \Box(p \wedge q)?; \Box \neg r^u]\!]$. We write this out to check its meaning:

$$s[\![\Box p^u; \Box(p \wedge q)?; \Box \neg r^u]\!]s'$$
$$\text{iff } \exists s'' : s[\![\Box p^u]\!]s'' \text{ and } s''[\![\Box(p \wedge q)?; \Box \neg r^u]\!]s'$$
$$\text{iff } \exists s'' : s[\![\Box p^u]\!]s'' \text{ and } s'' \models \Box(p \wedge q) \text{ and } s''[\![\Box \neg r^u]\!]s'$$
$$\text{iff } \exists s'' : s[\![\Box p^u]\!]s'' \text{ and } s'' \models \Box q \text{ and } s''[\![\Box \neg r^u]\!]s'.$$

It follows from this that the non-abort condition on s is given by:

$$\exists s' : s[\![\Box p^u]\!]s' \text{ and } s' \models \Box q.$$

This is the case iff $s \models \Box(p \rightarrow q)$. In other words, the presupposition is that it is known in the current context that if Jan is male then he is adult. We can also derive this in the calculus, as follows:

$$\langle \Box p^u; \Box(p \wedge q)?; \Box \neg r^u \rangle \top \quad \leftrightarrow \quad \langle \Box p^u \rangle \langle \Box(p \wedge q)? \rangle \langle \Box \neg r^u \rangle \top$$
$$\leftrightarrow \quad \langle \Box p^u \rangle \Box(p \wedge q)$$
$$\leftrightarrow \quad (\Box(p \wedge q))^p$$
$$\leftrightarrow \quad \Box(p \rightarrow q).$$

An information transition π *holds* in a context if the transition does not affect that context. For the example case, we can spell out the conditions for this as follows:

$$s[\![\Box p^u; \Box(p \wedge q)?; \Box \neg r^u]\!]s$$
$$\text{iff } \exists s' : s[\![\Box p^u]\!]s' \text{ and } s'[\![\Box(p \wedge q)?; \Box \neg r^u]\!]s$$
$$\text{iff } s[\![\Box p^u]\!]s \text{ and } s[\![\Box(p \wedge q)?; \Box \neg r^u]\!]s$$
$$\text{iff } s \models \Box p \text{ and } s \models \Box(p \wedge q) \text{ and } s \models \Box \neg r$$
$$\text{iff } s \models \Box(p \wedge q \wedge \neg r).$$

To end this section, note that the present perspective sheds an illuminating light on the phenomenon known as presupposition accommodation. Presupposition accommodation is the process performed by a benevolent audience in case an assertion is made with a presupposition which does not hold in the current context. In case the audience does not know that Bill is married and someone gossips that Bill's wife wants a divorce then the context is tacitly updated with the presupposition of that assertion as well. If we allow complex updates (by an obvious extension of the language), we can model this accommodation process as a shift from transition π to transition $(\langle \pi \rangle \top)^u; \pi$.

8 Embedded Presuppositions

Until now we have only considered presuppositions under sequential composition. If we assume presuppositions to have the form $\Box F?$, then a typical sequential composition of two updates under presupposition looks like this:

$$\Box F_1?; \Box G_1^u; \Box F_2?; \Box G_2^u.$$

The presupposition of this is given by:

$$\langle \Box F_1?; \Box G_1^u; \Box F_2?; \Box G_2^u \rangle \top.$$

This reduces to:

$$\Box F_1 \wedge \langle \Box G_1^u \rangle \langle \Box F_2? \rangle \top,$$

and further to:

$$\Box F_1 \wedge \Box (G_1 \to F_2),$$

with end result:

$$\Box (F_1 \wedge (G_1 \to F_2)).$$

Thus, we see that the boxed presupposition of the sequential composition of two updates is given by conjunction of the boxed presupposition of the first and the boxed implication of assertion of the first and presupposition of the second.

To consider presuppositions under negation, let us forget about 'downward' transitions π for the moment. (Note that we cannot define the negation of a downdate with α as the assertion that downdating with α itself would lead to inconsistency, for if a downdate with α is possible at all, it will never lead to inconsistency. Also, the negation of a downdate with α cannot be construed as the assertion that downdating with α is impossible, for the impossibility of a downdate with α just means that α is a logical truth.)

An *update transition* is a transition π with the property that $s[\![\pi]\!]s'$ implies $s \sqsubseteq s'$. The *presupposition* of an update transition is the set $\{s \in S \mid \exists s' \sqsupseteq s : s[\![\pi]\!]s'\}$. This set is characterized by $\langle \pi \rangle \top$. The *content* of an update transition is the set $\{s \in S \mid s[\![\pi]\!]s\}$. This set is characterized by fix (π).

Negating an update α^u can be construed as updating with the assertion that making update α itself would yield inconsistency. Thus, we can stipulate:

$$\neg(\alpha^u) = ([\alpha^u] \Box \bot)^u.$$

If we define $\varphi_1^u \Rightarrow \varphi_2^u$ as $\neg(\varphi_1^u; \neg\varphi_2^u)$ (update implication) and $\varphi_1^u \sqcup \varphi_2^u$ as $\neg(\neg\varphi_1^u; \neg\varphi_2^u)$ (update disjunction), then we can easily derive:

- $[\![\neg(\Box F^u)]\!] = [\![\Box \neg F^u]\!]$,
- $[\![\Box F^u \Rightarrow \Box G^u]\!] = [\![\Box(F \to G)^u]\!]$,
- $[\![\Box F^u \sqcup \Box G^u]\!] = [\![\Box(F \vee G)^u]\!]$.

In the general case where an update transition π may have a presupposition, we have two options: either the negation preserves the presupposition or it cancels it. We will explore the first option. Suppose π is an upward transition. Then we define $\neg\pi$ as follows:

$$\neg\pi \stackrel{\text{def}}{=} \langle \pi \rangle \top?; ([\pi] \Box \bot)^u.$$

Thus, $\neg\pi$ has the same presupposition as π, but it updates to the minimal state(s) where updating with π would yield inconsistency.

As regards the second option, an obvious choice for a definition of negated updating which cancels presuppositions is $([\pi]\Box\bot)^u$. This is an update to a state where doing π itself would lead to inconsistency. Now suppose π has a presupposition, let us say $\Box p?$, and assume $\Diamond\neg p$ holds in the current state. Then $([\pi]\Box\bot)^u$ would loop in the current state, showing that $([\pi]\Box\bot)^u$ does not have $\Box p?$ as presupposition.

If we spell out the semantics for $\neg\pi$ we get this:

$$[\![\neg\pi]\!] = A - B,$$

where

$$\begin{aligned} A \;=\; &\{\langle s,s'\rangle \in S \times S \mid s \sqsubseteq s', s'[\![\pi]\!]\langle\emptyset,w\rangle \text{ (for some } w),\\ &\text{and for all } s'' \text{ with } s\sqsubseteq s'' \sqsubseteq s' \;\&\; s''[\![\pi]\!]\langle\emptyset,w\rangle \text{ (for some } w)\\ &\hspace{8cm} s' \sqsubseteq s''\}, \end{aligned}$$

and

$$B = \{\langle s,s'\rangle \mid s \sqsubseteq s', s \models [\pi]\bot\}.$$

Note that the earlier stipulation for $\neg(\alpha^u)$ is a special case of this.

We can now define dynamic implication and dynamic disjunction for the general case of update transitions π_1 and π_2. To calculate what happens to presuppositions under 'dynamic implication', we can make use of the fact that

$$\neg(\Box F?; \Box G^u) = \Box F?; \Box\neg G^u$$

and of the fact that for all π:

$$\langle\neg\pi\rangle\top \leftrightarrow \langle\pi\rangle\top.$$

This is just a reflection of the fact that $\neg\pi$ has the same presupposition as π.

Here is the calculation for presupposition under dynamic implication:

$$\begin{aligned} &\langle(\Box F_1?; \Box G_1^u) \Rightarrow (\Box F_2?; \Box G_2^u)\rangle\top\\ \leftrightarrow\; &\langle\neg((\Box F_1?; \Box G_1^u); \neg(\Box F_2?; \Box G_2^u))\rangle\top\\ \leftrightarrow\; &\langle\neg(\Box F_1?; \Box G_1^u; \Box F_2?; \Box\neg G_2^u)\rangle\top\\ \leftrightarrow\; &\langle\Box F_1?; \Box G_1^u; \Box F_2?; \Box\neg G_2^u\rangle\top\\ \leftrightarrow\; &\Box F_1 \wedge \langle\Box G_1^u\rangle\Box F_2\\ \leftrightarrow\; &\Box(F_1 \wedge (G_1 \to F_2)). \end{aligned}$$

For presupposition under 'dynamic disjunction' we get:

$$\begin{aligned} &\langle(\Box F_1?; \Box G_1^u) \sqcup (\Box F_2?; \Box G_2^u)\rangle\top\\ \leftrightarrow\; &\langle\neg(\neg(\Box F_1?; \Box G_1^u); \neg(\Box F_2?; \Box G_2^u))\rangle\top\\ \leftrightarrow\; &\langle\neg(\Box F_1?; \Box\neg G_1^u; \Box F_2?; \Box\neg G_2^u)\rangle\top\\ \leftrightarrow\; &\langle\Box F_1?; \Box\neg G_1^u; \Box F_2?; \Box\neg G_2^u\rangle\top\\ \leftrightarrow\; &\Box F_1 \wedge \langle\Box\neg G_1^u\rangle\Box F_2\\ \leftrightarrow\; &\Box(F_1 \wedge (G_1 \vee F_2)). \end{aligned}$$

Thus, calculating presuppositions of complex updates in terms of assertions and presuppositions of their components gives the following table:

update procedure	presupposition
$(\Box F_1?;\Box G_1^u);(\Box F_2?;\Box G_1^u)$	$\Box(F_1 \wedge (G_1 \to F_2))$.
$\neg(\Box F?;\Box G^u)$	$\Box F$.
$(\Box F_1?;\Box G_1^u) \Rightarrow (\Box F_2?;\Box G_1^u)$	$\Box(F_1 \wedge (G_1 \to F_2))$.
$(\Box F_1?;\Box G_1^u) \sqcup (\Box F_2?;\Box G_2^u)$	$\Box(F_1 \wedge (G_1 \vee F_2))$.

This is a *boxed* version of Karttunen's table of presupposition projection for 'and', 'not', 'if then' and 'or'.

9   Digression: Error States

The treatment of presupposition failure in terms of error states of van Eijck (1993) and van Eijck (1994) is motivated by an obvious parallel between presupposition failure in natural language and error abortion in imperative programming. Consider the program statement (7).

(7)   x := y/z

If at the point of execution of this statement register z happens to contain the value 0 then execution will be aborted with an error statement like 'Floating point error: division by zero attempted'.

(8)   IF z <> 0 THEN x := y/z

In the statement (8) the dangerous case of $z = 0$ is tested for in the program code, and the danger of error abortion is staved off.

This suggests analyzing presupposition failure as 'moving to an error state'. Taking error abortion into account in the semantics of deterministic imperative programming boils down to changing the semantic interpretation function for program statements into a partial function: error abortion is the case where there is no next state.

The epistemic state of a program always consists of the current memory state, so it turns out that error abortion analysis arises as a special case of the present epistemic analysis, where there are just two state sets: $\langle\{w\}, w\rangle$ (the consistent state) and $\langle\emptyset, w\rangle$ (the inconsistent state). Thus we get:

Success   case $w(p) = 1, w(q) = 1$:

$$\langle\{w\}, w\rangle[\![\Box p?;\Box q^u]\!]\langle\{w\}, w\rangle$$

Failure   case $w(p) = 1, w(q) = 0$:

$$\langle\{w\}, w\rangle[\![\Box p?;\Box q^u]\!]\langle\emptyset, w\rangle$$

Error abortion   case $w(p) = 0$:

$$\langle\{w\}, w\rangle[\![\Box p?;\Box q^u]\!] \text{ ERROR } .$$

In nondeterministic imperative programming, program statements are interpreted as relations. Taking error abortion into account here means changing the interpretation relation into a partial relation. Executing a program statement π in state s now gives three possibilities: (1) there are proper next states (and maybe the program can also make a transition to 'error'), (2) there are no next states, and (3) the program can only make a transition to the error state. Again, an error state semantics for dynamic predicate logic (Groenendijk and Stokhof 1991) in the style of van Eijck (1993) turns out to be a special case of the present epistemic analysis, where there is just one first order model around, and where the possible states are the assignment functions over this single model (intuitively, the states encode the interpretations for indefinite noun phrases that are 'still in the running'). In this set-up, an update with presupposition for 'the king of France is bald' could be rendered as $\Box \exists !xKx?; \Box(x :=?; Kx; Bx)^u$. This gets us into the topic of the next section.

10 Presupposition and Quantification

As an example of a presupposition of quantified expressions, we look at the case of uniqueness presuppositions of singular definite descriptions. Dynamic versions of predicate logic have been proposed to deal with growth of information about anaphoric possibilities of a piece of natural language text. The most important ones of these are file change semantics (Heim 1982), discourse representation theory (Kamp 1981), and dynamic predicate logic (Groenendijk and Stokhof 1991). This kind of dynamics can be, but need not be, combined with the dynamics of information updating using predicate logical formulas. Here we will concentrate on 'epistemic dynamics' for purposes of exposition, and sketch a system of information updating for standard predicate logic.

To model information growth in predicate logic, the simplest possible set-up confines attention to one particular predicate logical model M for the language under consideration, and then uses sets of variable assignments for that model as information states. Thus, if $M = \langle \text{dom}(M), \text{int}(M) \rangle$ is given, and if V is the set of variables for the predicate logical language under consideration, then $A = \text{dom}(M)^V$ is the set of assignments, and $S = \{\langle i, a \rangle \mid i \subseteq A, a \in A\}$ is the set of information states. The relation $\sqsubseteq$ on S is given by $s \sqsubseteq s'$ iff $s = \langle i, a \rangle, s' = \langle j, a \rangle$ and $i \supseteq j$. Absurd information states are states of the form $\langle \emptyset, a \rangle$.

Fix a language L: let a set of individual constants C and a set of predicate constants P_n (where n denotes the arity of the constant) be given. Assume V is a set of individual variables. Assume $c \in C, v \in V$,

$R \in P_n$.

$$t \quad ::= \quad c \mid v$$
$$\varphi \quad ::= \quad \perp \mid Rt_1 \cdots t_n \mid t_1 = t_2 \mid \neg\varphi \mid (\varphi_1 \wedge \varphi_2) \mid \exists v\varphi.$$

Let L_1 be the language that allows epistemic statements over L:

$$\psi \quad ::= \quad \varphi \mid \Diamond\varphi \mid \Box\varphi.$$

The Tarskian satisfaction relation $M \models_a \varphi$ is defined in the usual manner. In terms of this we define an interpretation for L_1 (i.e., a specification function σ) as follows:

$$\sigma(\varphi) \quad = \quad \{s \in S \mid s = \langle i, a \rangle \text{ and } M \models_a \varphi\},$$
$$\sigma(\Diamond\varphi) \quad = \quad \{\langle i, a \rangle \in S \mid \exists a' \in i \text{ with } M \models_{a'} \varphi\},$$
$$\sigma(\Box\varphi) \quad = \quad \{\langle i, a \rangle \in S \mid \forall a' \in i : M \models_{a'} \varphi\}.$$

The dynamically extended language now becomes:

$$t \quad ::= \quad c \mid v$$
$$\varphi \quad ::= \quad \perp \mid Rt_1 \cdots t_n \mid t_1 = t_2 \mid \neg\varphi \mid (\varphi_1 \wedge \varphi_2) \mid \exists v\varphi$$
$$\psi \quad ::= \quad \varphi \mid \Diamond\varphi \mid \Box\varphi \mid \mathrm{dom}\,(\pi) \mid \mathrm{ran}\,(\pi) \mid \mathrm{fix}\,(\pi).$$
$$\pi \quad ::= \quad \psi^u \mid \psi? \mid (\pi_1; \pi_2).$$

Minimal updates are defined as before. The definitions of the dynamic operators are the same as before. Again, presuppositions are given by $\mathrm{dom}\,(\pi)$ and assertions by $\mathrm{fix}\,(\pi)$. The distinction between presupposition failure and updating with a piece of information inconsistent with the current information state is given by 'no further transition possible' versus 'transition to an absurd state'.

(9)   The king of France is eating a frog.

An update with the information expressed by (9) is expressed in this format as (10).

(10) $\Box\exists!xKx?; \Box\exists x(Kx \wedge \exists y(Fy \wedge Exy))^u$.

The presupposition is given by:

(11) $\langle\Box\exists!xKx?; \Box\exists x(Kx \wedge \exists y(Fy \wedge Exy))^u\rangle\top$.

This is equivalent to:

(12) $\Box\exists!xKx$.

The assertion is given by:

(13) $\mathrm{fix}\,(\Box\exists!xKx?; \Box\exists x(Kx \wedge \exists y(Fy \wedge Exy))^u)$.

This is equivalent to:

(14) $\Box(\exists!xKx \wedge \exists x(Kx \wedge \exists y(Fy \wedge Exy)))$.

Note that because our epistemic states are based on a single first order model the epistemic operators $\Diamond$ and $\Box$ are not very expressive. They serve to make the distinction between being able to make an update to an absurd state (uttering a falsehood) and not being able to make a further transition at all (error abortion). Indeed, since possible worlds are variable assignments, if F is a predicate logical formula without free variables, then the difference between $\Box F$ and $\Diamond F$ shows up only in absurd information states.

Of course, the epistemic modalities become more expressive once we redefine our information states in terms of *sets* of first order models.

11 Conclusion

We have sketched a system of epistemic dynamic logic to model presupposition and presupposition failure. Lots of logical questions remain to be answered. For instance: is the logic of propositional up- and downdating decidable? We conjecture that it is. What does a complete axiomatisation of this logic look like? What are the properties of the systems one gets by imposing further conditions on the information structures? What do the obvious variations on the combination of presupposition and quantification look like? The simplest variation is to replace standard predicate logic by dynamic predicate logic. This yields the dynamic error state semantics of van Eijck (1993). Another variation is to replace states based on single first order models by states based on sets of models. This gives an epistemic first order update logic. Finally, we can combine the two in various ways (see Eijck and Cepparello 1994 and Groenendijk, Stokhof and Veltman 1994). In all cases, the main thing is to get at the right definition of the information structure $\langle S, \sqsubseteq \rangle$. Jaspars and Krahmer (1995) provide a very useful starting point for this in the form of an overview of current systems of dynamic logic from the perspective of information structures.

References

Beaver, D.I. 1992. The Kinematics of Presupposition. In *Proceedings of the Eighth Amsterdam Colloquium*, ed. P. Dekker and M. Stokhof, 17–36. ILLC, University of Amsterdam.

Benthem, J. van. 1991. Logic and the Flow of Information. Technical Report LP-91-10. ILLC, University of Amsterdam.

Chellas, B.F. 1980. *Modal Logic: An Introduction*. Cambridge University Press.

Eijck, J. van. 1993. The Dynamics of Description. *Journal of Semantics* 10:239–267.

Eijck, J. van. 1994. Presupposition Failure — A Comedy of Errors. *Formal Aspects of Computing* 6A:766–787.

Eijck, J. van, and G. Cepparello. 1994. Dynamic Modal Predicate Logic. In *Dynamics, Polarity and Quantification*, ed. M. Kanazawa and C.J. Piñón. 251–276. CSLI, Stanford.

Eijck, J. van, and F.J. de Vries. 1995. Reasoning About Update Logic. *Journal of Philosophical Logic* 24:19–45.

Groenendijk, J., and M. Stokhof. 1991. Dynamic Predicate Logic. *Linguistics and Philosophy* 14:39–100.

Groenendijk, J., M. Stokhof, and F. Veltman. June, 1994. Coreference and Modality. Manuscript, ILLC, Amsterdam.

Halpern, J., and Y. Moses. 1985. Towards a Theory of Knowledge and Ignorance: Preliminary Report. In *Logics and Models of Concurrent Systems*, ed. K.R. Apt. 459–476. Springer.

Heim, I. 1982. *The Semantics of Definite and Indefinite Noun Phrases*. Doctoral dissertation, University of Massachusetts, Amherst.

Heim, I. 1983. On the Projection Problem for Presuppositions. *Proceedings of the West Coast Conference on Formal Linguistics* 2:114–126.

Heim, I. 1992. Presupposition Projection and the Semantics of the Attitude Verbs. *Journal of Semantics* 9(3):183–221. Special Issue: Presupposition, Part 1.

Jaspars, J. to appear. Partial Up and Down Logic. *Notre Dame Journal of Formal Logic*.

Jaspars, J., and E. Krahmer. 1995. Unified Dynamics. Manuscript, CWI, Amsterdam.

Kamp, H. 1981. A Theory of Truth and Semantic Representation. In *Formal Methods in the Study of Language*, ed. J. Groenendijk et al. Mathematisch Centrum, Amsterdam.

Kanazawa, M. 1994. Completeness and Decidability of the Mixed Style of Inference with Composition. In *Proceedings 9th Amsterdam Colloquium*, ed. P. Dekker and M. Stokhof. 377–390. ILLC, Amsterdam.

Karttunen, L. 1973. Presuppositions of Compound Sentences. *Linguistic Inquiry* 4:169–193.

Karttunen, L. 1974. Presupposition and Linguistic Context. *Theoretical Linguistics* 181–194.

Karttunen, L., and S. Peters. 1979. Conventional Implicature. In *Syntax and Semantics 11: Presupposition*, ed. C.-K. Oh and D. Dinneen. 1–56. Academic Press.

Krahmer, E. 1994. Partiality and dynamics; theory and application. In *Proceedings 9th Amsterdam Colloquium*, ed. P. Dekker and M. Stokhof. 391–410. ILLC, Amsterdam.

Moore, R.C. 1984. Possible world semantics for autoepistemic logic. In *Proceedings AAAI Workshop on Non-Monotonic Reasoning*, 344–354. New Paltz, NY.

Rijke, M. de. 1994. Meeting Some Neighbours. In *Logic and Information Flow*, ed. J. van Eijck and A. Visser. 170–195. MIT Press, Cambridge, Mass.

Stalnaker, R. 1972. Pragmatics. In *Semantics of Natural Language*, ed. D. Davidson and G. Harman. 380–397. Reidel.

Stalnaker, R. 1974. Pragmatic Presuppositions. In *Semantics and Philosophy*, ed. M.K. Munitz and P.K. Unger. 197–213. New York University Press.

Veltman, F. 1991. Defaults in Update Semantics. Technical report. Department of Philosophy, University of Amsterdam. To appear in the *Journal of Philosophical Logic*.

Zeevat, H. 1992. Presupposition and Accommodation in Update Semantics. *Journal of Semantics* 9(4):379–412. Special Issue: Presupposition, Part 2.

5

Indefeasible Semantics and Defeasible Pragmatics

MEGUMI KAMEYAMA

1 Introduction

An account of utterance interpretation in discourse needs to face the issue of how the discourse context controls the space of interacting preferences. Assuming a discourse processing architecture that distinguishes the grammar and pragmatics subsystems in terms of monotonic and nonmonotonic inferences, I will discuss how independently motivated default preferences interact in the interpretation of intersentential pronominal anaphora.

In the framework of a general discourse processing model that integrates both the grammar and pragmatics subsystems, I will propose a fine structure of the preferential interpretation in pragmatics in terms of defeasible rule interactions. The pronoun interpretation preferences that serve as the empirical ground draw from the survey data specifically obtained for the present purpose.

I would like to thank David Beaver, Johan van Benthem, Paul Dekker, Jan van Eijck, Jan Jaspars, Aravind Joshi, Alex Lascarides, Daniel Marcu, Becky Passonneau, Henriëtte de Swart, and Frank Veltman for helpful discussions and comments on earlier versions of the paper. The thoughtful comments by an anonymous reviewer helped reshape the focus of the paper. I also profited from the comments from the seminar participants at the University of Bielefeld and the University of Amsterdam. I would also like to thank those who responded to the pronoun interpretation questionnaire whose results are discussed herein. Part of the work was sponsored by project NF 102/62–356 ('Structural and Semantic Parallels in Natural Languages and Programming Languages'), funded by the Netherlands Organization for the Advancement of Research (N.W.O.).

Quantifiers, Deduction, and Context
Makoto Kanazawa, Christopher Piñón, and Henriëtte de Swart, editors
Copyright © 1996, CSLI Publications

2 Discourse Processing Architecture

I will assume in this paper that a *discourse* is a sequence of utterances produced (spoken or written) by one or more discourse participants. *Utterances* are tokens of sentences or sentence fragments with which the speakers communicate certain information, and it is done in a *context*. Utterance interpretation depends on the context, and utterance meaning updates the context.

A specification of the complex interdependencies involved in utterance interpretation is greatly facilitated if it is couched in a discourse processing architecture that is both logically coherent and as closely as possible an approximation of the human cognitive architecture for discourse processing. What are the major modules of the architecture, and what types of inferences do they support? I claim that the most fundamental separation is between the spaces of *possibilities* and *preferences*.

2.1 Separating Combinatorics and Preferences

There is an assumption in computational linguistics that combinatorics should take precedence over preferences. The wisdom is to maximize the combinatoric space of utterance interpretation and to keep a firm line between this space and the other, preferential, space of interpretation. Preferences are affected by computationally expensive open-ended commonsense inferences. Combinatorics determine all and only possible interpretations, and preferences prioritize the possibilities.[1] Seen from another point of view, combinatorics are *indefeasible* — that is, never overridden by commonsense plausibility, whereas preferences are *defeasible* — that is, can be overridden by commonsense plausibility. I will henceforth assume that the grammar subsystem consists only of indefeasible possibilities, hence monotonic, whereas the pragmatics subsystem consists mostly (or possibly entirely) of defeasible preferences, hence nonmonotonic.[2]

An example of indefeasible rules of grammar in English is the Subject-Verb-Object constituent order. The sentence *Coffee drinks*

---

[1] This separation of rule types does not imply a sequential ordering between the two processing modules. Different rule types can be interleaved for interpreting or generating a subsentential constituent.

[2] The same formal system can be viewed from different viewpoints — as a system of *rules*, *constraints*, or *inferences*. Rules produce and transform structures in a system, constraints reduce possible structures, and inferences are used to reason about structures (e.g., manipulating assertions or drawing conclusions) as the "logic" in the standard sense. To take a prominent example, in the "parsing as deduction" paradigm (Pereira and Warren, 1980), context-free rules are also seen as deductive inference rules. The rule $S \rightarrow NP\ VP$ is translated into the inference rule $NP(i,j) \wedge VP(j,k) \rightarrow S(i,k)$. I will not adhere to one particular viewpoint in this paper, and rather take advantage of the flexibility.

Sally uttered in a normal intonation cannot mean "Sally drinks coffee" despite the commonsense support. An example of defeasible preferences is the interpretation of the pronoun *he* in discourse "*John hit Bill. He was severely injured.*" The combinatoric rule of pronoun interpretation would say that both John and Bill are possible referents of *he*, while the preferential rule would say that Bill is preferred here because it is more plausible that the one who is hit gets injured rather than vice versa. Crucially, this preference is overridden in certain contexts. For instance, if Bill is an indestructible cyborg, the preferred semantic value of *he* would shift to John.

The inferential properties of the *grammar* subsystem as a space of possibilities are well–illustrated in the so-called unification–based grammatical formalisms (UBG). A UBG system consists of context-free phrase structure constraints and unification constraints. Maxwell and Kaplan (1993) describe how the constraint interactions can be made efficient by exploiting the following properties of a UBG system: (1) *monotonicity* — no deduction is ever retracted when new constraints are added, (2) *independence* — no new constraints can be deduced when two systems are conjoined, (3) *conciseness* — the size of the system is a polynomial function of the input that it was derived from, and (4) *order invariance* — sets of constraints can be processed in any order without changing the final result.[3]

The inferential properties of the *pragmatics* subsystem are much less understood. Its general features can be characterized as those of *preferential reasoning*, a topic more studied in AI than in linguistics. The pragmatics subsystem contains sets of preference rules that, in certain combinations, could lead to conflicting preferences. This fundamental indeterminacy leads to the properties opposite from those of the grammar subsystem: (1) *nonmonotonicity* — preferences can be canceled when overriding preferences are added, (2) *dependence* — new preferences may result when two pragmatic subsystems are conjoined, (3) *explosion* — the system size is possibly an exponential (or worse) function of the input that it was derived from, and (4) *order variance* — changing the order in which sets of preferences are processed may also change the final result. The key to a discourse processing architecture is to preserve the above computational properties of the grammar subsystem while striving for a maximal control of the preference interactions in the pragmatics subsystem.[4]

---

[3]Grammar rules can be seen from two viewpoints — they *eliminate* as well as *create* possibilities. The former applies when communication is seen as incremental elimination of possible information states. The latter applies when it is seen as incremental increase of information content. I leave the choice open here.

[4]In contrast, the abduction–based system (Hobbs et al., 1993) does not separate grammar and pragmatics. All the rules are defeasible and directly interact in one big

Existing logical semantic theories employing dynamic interpretation rules (e.g., Kamp, 1981; Heim, 1982; Groenendijk and Stokhof, 1991; Kamp and Reyle, 1993) formalize the basic context dependence of indefeasible semantics. While these theories predict the *possible* dynamic interpretations of utterances, they are not concerned with how to compute the relative preferences among them. Lascarides and Asher (1993) extend the Discourse Representation Theory (DRT) (Kamp, 1981) with the interaction of defeasible rules for integrating a new utterance content into the discourse information state. The input to their defeasible reasoning is a fully interpreted DR Structure (DRS), with all the NPs already interpreted. The pragmatics subsystem I am concerned with here also includes the defeasible rules for NP interpretation and constituent attachments needed for DRS construction. The input to pragmatics in the present proposal is a much less specified logical form, and pragmatics kicks in *during* DRS construction.

2.2 The Processing Architecture

The discourse processing architecture that I will assume in the background of the remainder of this paper is this.[5]

- Let *discourse* be a sequence of utterances, $utt_1, \ldots, utt_n$. We say that utterance utt_i defines a *transition relation* between the *input context* C_{i-1} and the *output context* C_i. Context C is a multicomponent data structure (see section 2.3). The transition takes place as follows:

  - Let *grammar* G consist of rules of syntax and semantics that assign each utterance utt_i the *initial logical form* Φ_i.

  - Φ_i represents a disjunctive set of underspecified formulas containing unresolved references, unscoped quantifiers, and vague relations. Φ_i is the weakest formula that packages a *family* of formulas that covers the entire range of possible interpretations of utt_i (see section 3).

  - Let *pragmatics* P consist of rules for specifying and disambiguating Φ_i in context C_{i-1}. Ideally, P outputs the single *preferred interpretation* φ_i^k (φ_i^k is subsumed by Φ_i

---

module. (The defeasibility of grammar rules is motivated by the fact of disfluencies in language use.) The result is an increased computational complexity.

[5]This architecture is in line with Stalnaker's (1972:385) conception:

The syntactical and semantic rules for a language determine an interpreted sentence or clause; this, together with some features of the context of use of the sentence or clause, determines a truth value. An interpreted sentence, then, corresponds to a function from contexts into propositions, and a proposition is a function from possible worlds into truth values.

and there is no φ_i^j that is preferred over φ_i^k and also subsumed by Φ_i), and integrating φ_i^k into context C_{i-1} produces the *preferred output context* C_i. In a less felicitous case, the rules of P do not converge, resulting in multiple interpretations and output contexts.

2.3 Context

My aim here is to introduce the basic components of the context C in the above discourse processing architecture that I assume in the remainder of the paper.

Context C_i is a 6-tuple $\langle \varphi_i^k, D_i, A_i, I_i, L, K \rangle$ consisting of the fast-changing components, $\langle \varphi_i^k, D_i, A_i, I_i \rangle$, significantly affected by the dynamic import of utterances and the slow-changing components, $\langle L, K \rangle$, relatively stable in a given stretch of discourse instance. φ_i^k is the preferred interpretation (see section 2.2) of the last utterance utt_i in a logical form that preserves aspects of the syntactic structure of utt_i — best thought of as a short-term register of the surface structure of the previous utterance similar to the proposal by Sag and Hankamer (1984). D_i is the *discourse model* — a set of information states that the discourse has been about, which also incorporates the content of φ_i^k. D_i contains sets of situations, eventualities, entities, and relations among them, associated with the evolving event, temporal, and discourse structures. A_i is the *attentional state* — a partial order of the entities and propositions in D_i, where the ordering is by *salience*. A_i is separated from D_i because the same D_i may correspond to different variants of A_i depending on the particular sequence of utterances in particular forms describing the same set of facts. I_i is the set of *indexical anchors* — the indexically accessible objects in the current discourse situation — for instance, the values of indexical expressions such as *I, you, here,* and *now*. The slow-changing components are the *linguistic knowledge L* and *world knowledge K* used by the discourse participants. Although we know that discourse participants never share exactly the same mental state representing these components of the context, there must be a significant overlap in order for a discourse to be mutually intelligible. For the purpose of this paper, I will simply assume that context C is sufficiently shared by the participants.

The next section elaborates on the initial logical form Φ_i that plays a crucial role of defining the grammar–pragmatics boundary in the discourse processing architecture.

3 Indefeasible Semantics

The initial logical form (ILF) Φ represents the utterance's structure and meaning at the grammar–pragmatics boundary. This section discusses the general features of ILF with examples.

3.1 General Considerations

There are specific proposals for the ILF Φ in the computational literature (e.g., Alshawi and van Eijck, 1989; Alshawi, 1992; Alshawi and Crouch, 1992; Hwang and Schubert, 1992a, 1992b; Pereira and Pollack, 1991). Details in these proposals vary, but there is a remarkable agreement on the general features.

The ILF Φ contains "vague" predicates and functions representing *what* the utterance communicates. Vague predicates and functions represent various expression and construction types whose interpretation depends on the discourse context. They include unresolved referring expressions such as the pronoun *he*, unscoped quantifiers such as *each*, vague relations such as the relation between the nouns in a noun–noun compound, unresolved operators such as the tense operator *past* and the mood operator *imperative*, and attachment ambiguities such as for PP–attachments. The idea can also be extended to underspecify lexical senses at the ILF level. These predicates and functions generate 'assumptions' that need to be resolved or 'discharged' in the union of the discourse and sentence contexts. The ILF is thus *partial* and *indefeasible* — partial because it does not always have a truth value, and indefeasible because further contextual interpretations only prioritize possibilities and specify vagueness.

The ILF Φ also represents aspects of the utterance's surface structure relevant to *how* the utterance communicates the information content (e.g., the Topic–Focus Articulation of Sgall et al., 1986). Such a syntax–semantics corepresentation could be achieved in either of the two options: (1) the logical form is *structured*, representing aspects of phonological and surface syntactic structures such as the grammatical functions of nominal expressions, linear order, and topic–comment structure, or (2) the partial semantic representation and the phonological and syntactic structures are separately represented with mappings among corresponding parts. In this paper, the choice is arbitrary as long as certain syntactic information is available at the logical form.

There is a general question of *how far* and *how soon* the ILF gets specified and disambiguated by the pragmatics. The above existing proposals in the computational literature assume that each utterance is completely specified and disambiguated before the next utterance comes in. This includes the integration of the utterance content into the evolving discourse structure, event structure, and temporal structure in the context, as discussed by Lascarides and Asher (1993). An utterance's complete interpretation is not in general available on the spot, however, and it often has to wait till some more information is supplied in the subsequent discourse (Grosz et al., 1986). It is also possible that only the information concerning those entities that are significant or salient (or 'in focus') in the current discourse need to be fully specified and disambiguated.[6] The present discourse processing architecture allows such incremental and partial specification and disambiguation of the information state along discourse progression though this perspective is not explored in any technical detail here.

In sum, the ILF represents the indefeasible semantics of an utterance by leaving the following context–dependent interpretations underdetermined: reference of nominal expressions, modifier attachments, quantifier scoping, vague relations, and lexical senses. The ILF also leaves open how the given utterance is integrated into the temporal, event, and discourse structures in the context.

3.2   Our Working Formalism

I will use a simplified ILF in this paper. It is an underspecified predicate logic in a davidsonian style — a version of QLF (Kameyama, 1995) without the aterm–qterm distinction. The ILF for the utterance *"He made a robot spider"* is as follows:

$$decl\ (past[\exists exy[make(e) \land Agent_{Subj}(e, x) \land pro(x) \land he(x)$$
$$\land\,Theme_{Obj}(e, y) \land indef\_sg(y) \land spider(y)$$
$$\land\,nn\_relation(y, \lambda z(robot, z))]])$$

It contains the following vague predicates and functions:

- unresolved unstressed pronoun "he" — $pro(x) \land he(x)$

- unscoped quantificational determiner "a" — $indef\_sg(y)$

- a vague relation for a noun-noun compound "robot spider" — $spider(y) \land nn\_relation(y, \lambda z(robot, z))$ (a relation between a spider entity and a robot property)

- unresolved past tense — $past(\psi)$

- unresolved declarative mood — $decl(\psi')$

If the preferred interpretation of the utterance is that "John" made a robot shaped like a spider, we have the following DRS–like logical form:

$$\exists etxy[make(e) \land Time(e, t) \land Agent_{Subj}(e, x)$$
$$\land\,named(x, ``john") \land Theme_{Obj}(e, y) \land spider\_like(y) \land robot(y)]$$

The interpretation is complete when the content is integrated into the discourse, event, and temporal structures in the context. These struc-

---

[6]A comment by Paul Dekker.

tures are assumed to be in the discourse model D. The pragmatics sub-system must make all of the preferential decisions including NP inter-pretation and operator interpretation as well as contextual integration.[7]

3.3 Ambiguity and Underspecification

The initial logical form mixes both ambiguity and underspecification. The choice is largely arbitrary when the number of possible interpreta-tions is exhaustively enumerable. Whenever there are n possible inter-pretations for a linguistic item or construction type, we can have either (1) a disjunctive set of n interpretations $i_1, ..., i_n$, from which the prag-matics chooses the best, or (2) one underspecified interpretation that the pragmatics further specifies. Pragmatic disambiguation and speci-fication involve exactly the same kind of an interplay of linguistic and commonsense preferences, and relative preferences in disambiguation and specification are often interdependent.

Consider *He made a robot spider with six legs*. There is a preference for the interpretation "a robot spider with six legs" over the alternative "a male person with six legs". This preference is overridden in certain contexts — for instance, if the person is a fictional figure who can freely change the number of legs to be two, four, or six, the alternative reading becomes equally plausible. Note that the attachment disambiguation and pronoun interpretation are interdependent here.

When the number of possible interpretations cannot be exhaustively enumerated, however, ambiguity and underspecification are not inter-changeable, and we must posit an *underspecified relation* as a semantic primitive. A sufficient but not necessary condition for positing an un-derspecified relation is this (Kameyama, 1995):[8]

> An underspecified relation is posited when there is an open–ended set of possible specific relations associated with a construction type, and the interpretation is typically affected by *ad hoc* facts known in the discourse context.

A canonical example is the interpretation of noun–noun compounds such as *elephant pen*. It could mean a pen shaped like an elephant, a pen with elephant pictures on the body, a pen with a small toy elephant glued on the top, or, depending on the context, a pen that the speaker found on the ground when she was pretending to be an elephant. All we can tell from the grammar of noun-noun compounds is that it is a pen that has some salient relation with elephants. It makes sense, then,

---

[7]I assume that various preferential decisions are interleaved rather than sequen-tially ordered within pragmatics.

[8]We have here an operational criterion for separating out grammar and pragmat-ics. It leads to a discovery of cross–linguistic variation in the grammar–pragmatics boundary. Long–distance dependency is a case in point (Kameyama, 1995).

to explicitly state in the grammar output the vague notion of "some salient relation" as a primitive. This is the basic motivation of the proposal for underspecified relations in the logical form in the computational literature (e.g., Alshawi, 1990; Hobbs et al., 1993). The same thing goes with scope ambiguities. The number of possible scopings is always bounded but possibly very large (on the order of hundreds), and speakers are often unable to select a single specific scoping, so the grammar should defer assigning specific scopings to a sentence and give it to pragmatics (Hobbs, 1983; Reyle, 1993; Poesio, 1993).

In sum, with the ILF sealing off the space of grammatical reasoning, the present discourse processing architecture magnifies the importance of pragmatics in utterance interpretation. Pragmatics achieves anaphora resolution, attachment disambiguation, quantifier scoping, vague relation specification, and contextual integration all in one module. Is there a system in the chaos? That is the question we turn to now.

4 Defeasible Pragmatics

This section discusses the features and examples of the defeasible rules in the pragmatics subsystem.

4.1 General Considerations

By *defeasible*, I mean a conclusion that has to be retracted when some additional facts are introduced. This characterizes the *preferential* aspect of utterance interpretation with the nonmonotonicity property. Grammatical reasoning is governed by the Tarskian notion of valid inference in standard logic — "Each model of the premises is also a model for the conclusion." Pragmatic reasoning distinguishes among models as to their relevance or plausibility, and is governed by the notion of plausible inference (Shoham, 1988) — "Each *most preferred* model of the premises is a model for the conclusion." The preference can be stated in terms of default rules as well, so the general reasoning takes the form of "as long as no exception is known, prefer the default." In utterance interpretation, this form of reasoning chooses the best interpretation from among the set of possible ones. The present focus is the interpretation preferences of intersentential pronominal anaphora.

4.2 Earlier Computational Approaches to Pronoun Interpretation

Computational research on pronoun interpretation has always recognized the existence of powerful grammatical preferences, but there are different views on their status in the overall processing architecture. Hobbs (1978) discussed the relative merit of purely grammar–based and purely commonsense–based strategies for pronoun interpretation. His grammar–based strategy that accounts for 98% of a large number of pronouns in naturally occurring texts simply could not be extended to account for the remaining cases that only commonsense reasoning can explain. He settled in a "deeper" method that seeks a global *coherence* arguing that *coreference* can be determined as a side–effect of coherence–seeking interpretation. The abduction–based approach (Hobbs et al., 1993) is an example of such a general inference system, where syntax–based preferences for coreference resolution are used as the *last resort* when other inferences do not converge.

Sidner's (1983) local focusing model used an *attentional* representation level to mediate the grammar's *control* of discourse inferences. For each pronoun, there is an ordered list of potential referents determined by local focusing rules, and the highest one that leads to a consistent commonsense interpretation of the utterance is chosen. Common sense has a veto power over grammar-based focusing in the ultimate interpretation, but common sense *is* the last resort, contrary to Hobbs's approach. Carter (1987) implemented Sidner's theory combined with Wilks's (1975) preferential semantics, and reported the success rate of 93% for resolving pronouns in a variety of stories — of which only 12% relied on commonsense inferences.

Grammar's role in the control of inferences was the original motivation of the *centering model* (Joshi and Kuhn, 1979; Joshi and Weinstein, 1981). The proposal was to use the *monadic* tendency of discourse (i.e., tendency to be centrally about one thing at a time) to control the *amount of computation* required in discourse interpretation. Grosz, Joshi, and Weinstein (1983) proposed a refinement of Sidner's model in terms of centering, and highlighted the crucial role of pronouns in linking an utterance to the discourse context. Subsequent work on centering converged on an equally significant role of the main clause SUBJECT[9] (Kameyama, 1985, 1986; Grosz, Joshi, and Weinstein, 1986; Brennan, Friedman, and Pollard, 1987). Hudson D'Zurma (1988) experimentally verified that speakers had a difficulty in interpreting a discourse where a centering prediction was in conflict with commonsense plausibility, leading to a 'garden path' effect. An example from her experiment is: *"Dick had a jam session with Brad. He played trumpet while Brad played bass. ??He plucked very quickly."* Centering models the local attentional state management in an overall discourse model proposed by Grosz and Sidner (1986).

These computational approaches to discourse have recognized the non–truth–conditional effects on utterance interpretation coming from

---

[9]Grammatical functions will be in uppercase in order to avoid the ambiguity of these words.

the utterance's *surface structure* (i.e., phonological, morphological, and syntactic structures). Although this aspect of interpretation cannot be neglected in a discourse processing model, its relevance to a logical model of discourse semantics and pragmatics has remained unclear. It is worth pointing out that discourse pragmatics in the above computational approaches as well as in philosophy (e.g., Lewis, 1979; Stalnaker, 1980) has generally assumed a dynamic architecture. Would there be a potential fit with the dynamic semantic theories in linguistics (e.g., Kamp, 1981; Heim, 1982; Groenendijk and Stokhof, 1991) in a way that forms a basis for an integrated logical model of discourse semantics and pragmatics? In this paper, I propose a pronoun interpretation model taking ideas from *both* computational and linguistic traditions, and present it in such a way that it becomes tractable for logical implementation.

5 Pronoun Interpretation Preferences: Facts

Pronoun interpretation must be carried out in an often vast space of possibilities, somehow controlling the inferences with default preferences coming from different aspects of the current context. Pronouns such as *he, she, it* and *they* can refer to entities talked about in the current discourse, present in the current indexical context, or simply salient in the model of the world implicitly shared by the discourse participants. Since the problem space is vast and complex, we need to narrow it down to come to grips with interesting generalizations. I will now limit our discussion to the interpretation of the anaphoric use of *unstressed* male singular third person pronoun *he* or *him* in English.

5.1 Survey and the Results

In 1993, I conducted a survey of pronoun interpretation preferences using the discourse examples shown in Table 1. These examples were constructed to isolate the relevant dimensions of interest based on previous work (see section 5.2).

One set of examples, A–H, involves pronouns that occur in the second of two–sentence discourses. They were presented to competent (some nonnative) speakers of English in the A-F-C-H-E-D-B-G order, avoiding sequential effects of two adjacent similar examples. The speakers were instructed to read them with no special stresses on words, and to answer the who-did-what questions about pronouns in italics. The answer "unclear" was also allowed, in which case, the speaker was encouraged to state the reason. The total number of the speakers was 47, of which 10 were nonlinguist natural language researchers and 4 were nonnative but fluent English speakers. The second set of exam-

Grammatical Effects:
A.   John hit Bill. Mary told *him* to go home.
B.   Bill was hit by John. Mary told *him* to go home.
C.   John hit Bill. Mary hit *him* too.
D.   John hit Bill. *He* doesn't like *him*.
E.   John hit Bill. *He* hit *him* back.
K.   Babar went to a bakery. He greeted the baker. *He* pointed at a blueberry pie.
L.   Babar went to a bakery. The baker greeted him. *He* pointed at a blueberry pie.
Commonsense Effects:
F.   John hit Bill. *He* was severely injured.
G.   John hit Arnold Schwarzenegger. *He* was severely injured.
H.   John hit the Terminator. *He* was severely injured.
I.   Tommy came into the classroom. He saw Billy at the door. He hit him on the chin. *He* was severely injured.
J.   Tommy came into the classroom. He saw a group of boys at the door. He hit one of them on the chin. *He* was severely injured.

TABLE 1   Discourse Examples in the Survey

ples, I–L, are longer discourses. They were given to disjoint sets of native English speakers, none of whom are linguists.

The examples fall under two general categories, as indicated in Table 1. One group isolates the *grammatical effects* by minimizing commonsense biases. In these examples, it is conjectured that there is no relevant commonsense knowledge that affects the pronoun interpretation in question. The other group examines the *commonsense effects* of a specific causal knowledge of hitting and injuring in relation to the grammatical effects observed in the first group.

Table 2 shows the survey results. The $\chi^2_{df=1}$ significance for each example was computed by adding an evenly divided number of the "unclear" answers to each explicitly selected answer, reflecting the assumption that an "unclear" answer shows a genuine ambiguity. Preference is considered *significant* if $p < .05$, *weakly significant* if $.05 < p < .10$, and *insignificant* if $.10 < p$. Insignificant preference is interpreted to mean ambiguity or incoherence. It follows from the Gricean Maxim that ambiguity must be avoided in order for an utterance to be pragmatically felicitous. An example with an insignificant preference is thus infelicitous, and should not be generated.

It must be noted that the present survey results exhibit only one aspect of preferential interpretation — namely, the *final* preference reached after an unlimited time to think. They do not represent the *process* of interepretation — for instance, a number of speakers commented that they had to *retract* the first obvious choice in example I. This garden–path effect verified in Hudson D'Zurma's (1988) experiments does not show in the present survey results.

	Answers			$\chi^2_{df=1}$	p
A.	John 42	Bill 0	Unclear 5	37.53	$p < .001$
B.	John 7	Bill 33	Unclear 7	14.38	$p < .001$
C.	John 0	Bill 47	Unclear 0	47	$p < .001$
D.	J. dislikes B. 42	B. dislikes J. 0	Unclear 5	37.53	$p < .001$
E.	John hit Bill 2	Bill hit John 45	Unclear 0	39.34	$p < .001$
K.	Babar 13	Baker 0	Unclear 0	13	$p < .001$
L.	Babar 3	Baker 10	Unclear 0	3.77	$.05 < p < .10$
F.	John 0	Bill 46	Unclear 1	45.02	$p < .001$
G.	John 24	Arnold 13	Unclear 10	2.57	$.10 < p < .20$
H.	John 34	Terminator 6	Unclear 7	16.68	$p < .001$
I.	Tommy 3	Billy 17	Unclear 1	9.33	$.001 < p < .01$
J.	Tommy 10	Boy 7	Unclear 3	0.45	$.50 < p < .70$

TABLE 2 Survey Results

5.2 Discussion of the Results

The present set of examples highlights four major sources of preference in pronoun interpretation — *SUBJECT Antecedent Preference, Pronominal Chain Preference, Grammatical Parallelism Preference,* and *Commonsense Preference.* These are stated at a descriptive level with no theoretical commitments. A theoretical account of the same set of facts will be given in section 6. Each source of preference is discussed below.

SUBJECT Antecedent Preference. A hierarchy of the preferred intersentential antecedent of a pronoun has been proposed in the centering framework, which basically says that the main clause SUBJECT is preferred over the OBJECT (Kameyama, 1985,1986; Grosz et al., 1986). This preference is confirmed in examples A and B.[10]

The consistency of this preference across examples A and B demonstrates that grammatical functions rather than thematic roles are the adequate level of generalization. In both A and B, the thematic roles of Bill and John in the first sentence are agent and theme (or patient), respectively, but the switch in grammatical functions by passivization causes the preferred interpretation to switch accordingly.

Example C demonstrates the defeasibility of this preference in the face of the parallelism induced by the adverb *too* as a side effect of an indefeasible *conventional presupposition* (see section 6).

Pronominal Chain Preference. This is the preference for a chain

---

[10]Some speakers indicated that they had to assume additional facts in order to make a plausible scenario — for instance, in example A, "Mary is a teacher, and she sent John home as a punishment". The speakers seem to want some more information to make the judgment more conclusive. What are the relationships among these three people mentioned out of the blue? I realize that impoverished examples of this sort rarely occur in our real–life discourses. To sort out some rather delicate interplay of preferences, however, we need to start out with simplified examples. This is analogous to the use of the "blocks world" (i.e., the world of blocks) in AI.

of pronouns across utterances to corefer.[11] Examples K and L are a minimal pair of structural effects without a commonsense bias. Their contrast shows the effect of grammatical positions. The SUBJECT–SUBJECT chain of pronouns (example K) supports a significant coreference preference ($p < .001$), whereas the OBJECT-SUBJECT chain (example L) supports a weakly significant *noncoreference* preference ($.05 < p < .10$) indicating a parallelism effect below.

Example I shows that the causal knowledge also *in the end* overrides a stretch of SUBJECT pronominal chain, but as noted above, this example causes the speakers to first interpret the SUBJECT pronouns to corefer, then *retract* the choice due to the inconsistency with a causal knowledge. This processing tendency indicates that the grammatical preference is processed faster than the commonsense preference. We will come back to this issue later.

In example J, the strong preference for a SUBJECT pronominal chain is undermined by the indefiniteness of the referent (*one of the boys*) that the generic causal knowledge supports and by the additional inference — when one hits one of a group of boys, he would be revenged by the group. The grammar–based preference and common sense are in a tie here, showing a genuine ambiguity ($.50 < p < .70$).

Grammatical Parallelism Preference. There is a general preference for two adjacent utterances to be grammatically parallel. The parallelism requires, roughly, that the SUBJECTs of two adjacent utterances corefer and that the OBJECTs, if applicable, also corefer. This preference is demonstrated in example D that involves two pronouns.[12] In example L, the parallelism preference overrides the pronominal chain preference.

Example E shows the defeasibility of the parallelism preference in the face of the presupposition triggered by adverb *back*. An "x hit y back" event conventionally presupposes that a "y hit x" event has previously occurred, leading to the near-unanimous interpretation "Bill hit John back."[13]

Commonsense Preference. Examples F–H illustrate the effect

---

[11] I will use the simple terminology of "referent" and "coreference" without committing to their realist connotation because this does not affect the points I wish to make in this paper.

[12] Another possible source of preference is the *causal link* between the two described eventualities, John's hitting Bill ($e1$) and someone disliking someone ($e2$). The preferred interpretation supports the causal link "$e1$ *because* $e2$", while the alternative interpretation, which nobody took, supports "$e1$ *therefore* $e2$". These could be stated in terms of discourse relations of *Explanation* and *Cause* (e.g., Lascarides and Asher, 1993). I'm not aware of any empirical studies of this kind of preference effects.

[13] I suspect that the two speakers who took the opposite interpretation used the sense of *back* close to "again".

of a simple causal knowledge that dictates the final interpretation over and above the grammatical preferences. In example F, the SUBJECT Antecedent Preference is defeated by an inference derived from the generic causal knowledge — "when X hits Y, Y is normally hurt," and "being injured is being hurt." Since the example involves some "normal" fellows called John and Bill, it applies with full force (46/47).

Examples G and H show what happens to this baseline default when the described event involves some special individuals (fictitious or non-fictitious) that the speakers have some knowledge about. In example H, the preferred interpretation (34/47) swings to the one where the normal fellow, John, is injured as a result of attempting to assault the indestructible cyborg.[14] The cyborg also could have been injured (6/47) (because the movie showed that it *can* be destroyed after all). In example G, John attempts to assault a warm–blooded real person, Arnold, who seems a little stronger than normal fellows. Here, more speakers thought that John was injured (24/47) than Arnold was (13/47), but this preference is insignificant ($.10 < p < .20$). It reflects the indeterminacy of whether Arnold is a normal fellow or not, which affects the applicability of the generic causal knowledge.[15]

5.3 Descriptive Generalizations

Table 3 summarizes the preference predicted by each of the four sources discussed above and the final outcome verified in the survey. We see the following general patterns of conflict resolution:

1. Conventional Presuppositions (triggered by adverbs in examples C and E) and Commonsense Preferences (examples F, G, and H) dictate the *final* preference.
2. Grammatical Preferences take charge in the *absence* of relevant Commonsense Preferences (examples A–E, K, and L).
3. The SUBJECT Antecedent Preference overrides the Grammatical Parallelism Preference when in conflict (see examples A and B), and both are in turn stronger than the Pronominal Chain Preference (example L).

The cases of indeterminate final preference in examples G and J are worth noting. This kind of an indeterminate preference is infelicitous and uncooperative, which should be avoided in discourse generation. The indeterminacy in example G is due to the indeterminacy of Arnold being a normal person subject to injury or an abnormally strong person

---

[14]The Terminator is a cyborg played by Arnold Schwarzenegger in a popular science–fiction movie.

[15]Of interest here is the fact that the three speakers who knew *nothing* about what a "Terminator" is *all* interpreted that John was injured in example H. They clearly sensed "something nasty and abnormal" from this name alone.

	Subj.Pref.	Pron.Chain	Parallel.	Com.Sense	Outcome
A.	John	—	Bill	unclear	John
B.	Bill	—	John	unclear	Bill
C.	John	—	Bill	unclear	Bill♣
D.	John–Bill?	—	John–Bill	unclear	John–Bill
E.	John–Bill?	—	John–Bill	unclear	Bill–John◇
K.	Babar	Babar	Babar	unclear	Babar
L.	Baker	Babar	Baker	unclear	Baker
F.	John	—	John	Bill	Bill
G.	John	—	John	John/Arnold	John/Arnold
H.	John	—	John	John	John
I.	Tommy	Tommy	Tommy	Billy	Billy♠
J.	Tommy	Tommy	Tommy	Boy	Tommy/Boy

♣ — due to the conventional presupposition triggered by adverb *too*.
◇ — due to the conventional presupposition triggered by adverb *back*.
♠ — Tommy is the first choice, which is later retracted.

TABLE 3 Preference Interactions: Facts

who would not let himself be injured. The indeterminacy in example J is due to the conflict between the general causal knowledge about an injury caused by hitting and the insalience of an indefinite referent as a possible pronominal referent.

6 Pronoun Interpretation Preferences: Account

Four major sources of preference have been identified in the above pronoun interpretation examples. I propose that these sources correspond to the data structures in the different context components outlined in section 2.3. The context components the most relevant to the present discussion are the attentional state A, the LF register φ, and the discourse model D.

The main thrust of the present account is the general interaction of preferences that apply on different context components. It explains the basic fact that preferences may or may not be determinate. The present perspective of preference interactions also extends and explains the role of the attentional state in Grosz and Sidner's (1986) discourse theory.

6.1 The Role of the Attentional State

A discourse describes situations, eventualities, and entities, together with the relations among them. The attentional state A represents a dynamically updated snapshot of their *salience*. We thus assume the property *salient* to be a primitive representing the *partial order* among a set of entities in A.[16] The property *salient* is gradient and relative. A certain absolute degree of salience may not be achieved by any entities in a given A, but there is always a set of *maximally salient* entities,

---

[16]I will not discuss the partial order of propositions.

which is often, but not necessarily, a singleton set.[17] Thus it is crucial that a rule about the single maximally salient entity in a given A is only sometimes determinate.

We will now recast some elements of the centering model in the present discourse processing architecture. In the input context C_{i-1} for utterance utt_i, the form and content (φ_{i-1}) of the immediately preceding utterance utt_{i-1} occupy an especially salient status. The entities realized in utt_{i-1} are among the most salient subpart of A_{i-1}. I assume that this is achieved by a general A-updating mechanism. One of the entities in A_{i-1} may be the $Center_{i-1}$, what the current discourse is *centrally about*, hence the high salience:[18]

CENTER The Center is normally more salient than other entities in the same attentional state.

At least two default linguistic hierarchies are relevant to the dynamics of salience.[19] One is the *grammatical function hierarchy* (GF ORDER), and the other is the *nominal expression type hierarchy* (EXP ORDER). The GF ORDER in utt_i predicts the relative salience of entities in the *output* attentional state A_i whereas the EXP ORDER in utt_i predicts the relative salience of entities *assumed* in the *input* attentional state A_{i-1}.[20] EXP ORDER is also crucial to the management of the Center (EXP CENTER):

GF ORDER: Given a hierarchy, [SUBJECT > OBJECT > OBJECT2 > OTHERS], an entity realized by a higher ranked phrase is normally more salient in the output attentional state.

EXP ORDER: Given a hierarchy, [ZERO PRONOMINAL > PRONOUN > DEFINITE NP > INDEFINITE NP],[21] an entity realized by a higher–ranked expression type is normally more salient in the input attentional state.

EXP CENTER: An expression of the highest ranked type normally realizes the Center in the output attentional state.

EXP CENTER can be interpreted in two ways. One computes the "highest–ranked type" per utterance, sometimes allowing a non-

---

[17]Those entities that are "inaccessible" in the DRT sense do not participate in the salience ordering, or even if they do, they are below a certain minimal threshold of salience.

[18]In the centering model, the entities realized in φ_{i-1} are the "forward–looking centers" (Cf), and $Center_{i-1}$ is the "backward–looking center" (Cb).

[19]Consituents' linear ordering and animacy are also relevant.

[20]This order also approximates the relative salience of entities in the *output* attentional state, as demonstrated in part in example J.

[21]There is a pragmatic difference between stressed and unstressed pronouns, which should be accounted for by an independent treatment of stress — for example, in terms of a preference reversal function (Kameyama, 1994b). This paper concerns only unstressed pronouns.

pronominal expression type to output the Center. The other takes it to be fixed, namely, only the pronominals. The choice is empirical. In this paper, I will take the second interpretation.

Since matrix subjects and objects cannot be omitted in English,[22] the highest–ranked expression type is the (unstressed) pronoun (see Kameyama, 1985:Ch.1). From EXP ORDER, it follows that a pronoun *normally* realizes a *maximally salient entity* in the input attentional state. A pronoun can also realize a submaximally salient entity if this choice is supported by another overriding preference. The grammatical features of pronouns also constrain the range of possible referents — for instance, a *he*–type entity is a male agent. The maximal salience thus applies on the suitably restricted subset of the domain for each type of pronoun.

The interactions of the above defeasible rules — CENTER, GF ORDER, EXP ORDER, and EXP CENTER — account for various descriptive generalizations. First, the SUBJECT Antecedent Preference follows from GF ORDER and EXP ORDER — SUBJECT is the highest ranked GF in the first utterance, and a pronoun in the second utterance realizes the maximally salient entity in the input A. Second, the coreference and noncoreference preferences in pronominal chains are accounted for. The strong coreference preference for a SUBJECT–SUBJECT pronominal chain (example K) comes from the fact that a SUBJECT Center is the single maximally salient entity, which leads to a determinate preference. In contrast, an OBJECT Center competes with the SUBJECT non–Center for the maximal salience, which leads to an indeterminate preference based on salience alone (example L). The indeterminacy is resolved, to some extent, by the Grammatical Parallelism Preference (section 6.2).[23]

The center transition types of "establishing" and "chaining" (Kameyama, 1985,1986) result from the interactions of CENTER, EXP ORDER, and EXP CENTER.[24] The Center is "established" when a pronoun picks a salient non–Center in the input context and makes it the Center in the output context. It is "chained" when a pronoun picks the Center in the input context and makes it the Center in the output context. Examples A–H are thus concerned with Center–establishing pronouns, whereas examples I–L are concerned with Center–chaining

---

[22]Except in a telegraphic register.

[23]This notion of the single maximally salient entity corresponds to the "preferred center" Cp (Grosz et al., 1986) that is determined solely by the GF ORDER. The difference here is that it is determined by *both* the Center and GF ORDER, predicting an indeterminacy in certain cases.

[24]What I have previously called *retain* is now called *chain*. It covers both CONTINUE and RETAIN technically distinguished by Grosz et al. (1986) and Brennan et al. (1987).

pronouns. These transition types are not the primitives that directly drive preferences, however.

6.2   The Role of the LF Register

The grammatical parallelism of two adjacent utterances in discourse affects the preferred interpretation of pronouns (Kameyama, 1986), tense (Kameyama, Passonneau, and Poesio, 1993), and ellipsis (Pruest, 1992; Kehler, 1993). This general tendency warrants a separate statement. Parallelism is achieved, in the present account, by a computation on the pair of logical forms, one in the LF register in the context, and the other being interpreted.

PARA: The LF register in the input context and the ILF being interpreted seek maximal parallelism.[25]

The present perspective of rule interaction explains the "property–sharing" constraint on Center–chaining (Kameyama, 1986) as follows. GF ORDER, EXP ORDER, and PARA join forces to create a strong grammatical preference for SUBJECT–SUBJECT coreference (examples D,K). When they are in conflict, that is, when the maximally salient entity is not in a parallel position, PARA is defeated (examples A,B). When maximal salience is indeterminate, the parallelism preference affects the choice (example L), leading to a noncoreference preference for an OBJECT–SUBJECT pronominal chain.

6.3   The Role of the Discourse Model

The discourse model contains a set of information states about situations, eventualities, entities, and the relations among them. It also contains the evolving discourse structure, temporal structure, and event structure. Both linguistic semantics and commonsense preferences apply on the same discourse model.

Lexically Triggered Presuppositions. Adverbs *too* and *back* trigger conventional presuppositionsabout the input discourse model. These presuppositions are part of lexical semantics, thus indefeasible.

Adverb *too* triggers a presupposition that appears to seek parallelism between an utterance in the context and the utterance being interpreted. This is actually due to a general *similarity* presupposition associated with *too*. Consider each of the following utterances immediately preceding "John hit Bill too": "Mary hit Bill", "John hit Mary", "Mary kicked Bill", "John kicked Mary", "Mary hit Jane", and ?"John called Bill". What's construed as 'similar' in each case is a function of the particular utterance pair, and intuitively, preferred pairs sup-

---

[25]This statement is intentionally left vague. See Pruest's (1992) MSCD operation for a general definition of parallelism preference, and my property–sharing constraint (Kameyama, 1986) for a subcase relevant to pronoun interpretation.

port more similarities. Thus similarity comes in degrees, and a parallel interpretation is due to the preference for a maximal similarity.

Adverb *back* triggers a presupposition for a *reverse* parallelism. That is, the utterance "Bill hit John back" presupposes that it occurred after "John hit Bill".

Commonsense Knowledge. In contrast to the above rules that belong to the linguistic knowledge, the commonsense knowledge consists of all that an ordinary speaker knows about the world and life. Formalizing common sense is a major research goal of AI, where nonmonotonic reasoning has been intensively studied. My goal here is not to propose a new approach to commonsense reasoning but simply to highlight its interaction with linguistic pragmatics in the overall pragmatics subsystem. We know one thing for sure — there will be a relatively small number of linguistic pragmatic rules that systematically interact with an open–ended mass of commonsense rules. Since the linguistic rules can be seen to *control* commonsense inferences, our aim is to describe the former as fully as possible, and specify how the "control mechanism" works. The commonsense rules posited in connection to the examples in this paper are thus meant to be exemplary. There will be different rules for each new example and domain to be treated. The linguistic rules, however, should be stable across examples and domains.

The single powerful causal knowledge at work in our examples is that hitting may cause injury on the hittee but less likely on the hitter:

HIT: When an agent x hits an agent y, y is normally hurt.

The effects of the Terminator and Arnold indicate that the applicability of the HIT rule depends on the normality of the agents involved. Relevant knowledge includes things like: An agent is normally vulnerable, Arnold is a normal agent or an abnormally strong agent, and Terminator is an abnormally strong agent.

6.4 Account of the Rule Interactions

We now state the preference interaction patterns observed in Table 3 above. The SUBJECT Antecedent Preference and Pronominal Chain Preference result from CENTER, GF ORDER, EXP ORDER, and EXP CENTER. These are the defeasible *Attentional Rules* (ATT) stating the preferred attentional state transitions. The Grammatical Parallelism Preference is PARA. This is an example of the defeasible *LF Rules* (LF) stating the preferred LF transitions. Conventional presuppositions triggered by *too* and *back* are examples of the indefeasible *Semantic Rules* (SEM) in the grammar constraining the interpretation in the discourse model. The causal knowledge of hitting is HIT, with associated knowledge ETC about agents, Terminator, and Arnold. These

	ATT	LF	WK	SEM	Winner
A.	John	Bill	unclear	—	ATT
B.	Bill	John	unclear	—	ATT
C.	John	Bill	unclear	Bill	SEM
D.	John–Bill?	John–Bill	unclear	—	LF
E.	John–Bill?	John–Bill	unclear	Bill–John	SEM
K.	Babar	Babar	unclear	—	ATT+LF
L.	Baker/Babar	Baker	unclear	—	ATT+LF
F.	John	John	Bill	—	WK
G.	John	John	John/Arnold	—	WK
H.	John	John	John	—	WK
I.	Tommy	Tommy	Billy	—	WK (with difficulty)
J.	Tommy	Tommy	Boy(/Tommy)	—	??

Rules: ATT={CENTER, GF ORDER, EXP ORDER, EXP CENTER}, LF={PARA}, WK={HIT, ETC}, SEM={TOO, BACK}.

TABLE 4 Preference Interactions: Account

are examples of the defeasible *Commonsense Rules* (WK) stating the preferred discourse model. Table 4 identifies the rules that dominate the *final* interpretation in examples A–L.

General Features. The first distinction among these rules is defeasibility. The SEM rules are indefeasible whereas all other rules are defeasible. It is predicted that indefeasible rules override all defeasible rules, as verified in examples C and E.

What factor determines the interaction pattern among the defeasible rules? The three context components — discourse model D, attentional state A, and LF register φ — all have their preferred transitions. The D preference results from *proposition–level* (or "sentence–level") inferences *directly* determining the preferred model whereas the A and LF preferences result from *entity–level* (or "term–level") inferences only *indirectly* determining the preferred model. We have seen that proposition–level preferences, if applicable, generally override entity–level preferences, albeit with a varying degree of difficulty.

Take two examples: (1) "*John met Bill. He was injured.*" and (2) "*John hit Bill. He was injured.*" In (1), the ATT and LF preference that the pronoun refers to John indirectly leads to the preference that *John was injured*, which becomes the overall preference in the absence of relevant WK rules. In (2), relevant WK rules directly support a proposition–level preference, *Bill was injured*, which wins out (with a varying degree of difficulty). These "flows of preference" during an utterance interpretation are illustrated below:

(1) $[_S[_{NP} \, he]:\{\text{John}{>}\text{Bill}\} \; was \; injured] \implies$ John was injured

(2) $[_S[_{NP} \, he]:\{\text{John}{>}\text{Bill}\} \; was \; injured]:\{\text{Bill was injured} > \text{John was injured}\} \implies$ Bill was injured.

Conflict Resolution Patterns. We see a straightforward overriding pattern in examples A–H involving "Center–establishing" pronouns: *ATT* overrides *LF*, and *WK* overrides *ATT* and *LF*. Such an overriding relation can be seen as a dynamic updating operation (;) (van Benthem et al., 1993) — preferences are evaluated in turn, the later ones overriding the earlier ones: $LF; ATT; WK$.[26] It may be the general pattern of "changing preferences" during utterance interpretation.

Examples I–L involving "Center–chaining" pronouns show more or less the same pattern except that the overriding gets more difficult in some cases. It is more difficult when a SUBJECT pronoun chain supports a single maximally salient entity as in example I. This shows that the LF and ATT preferences in fact join forces to interact with the WK preferences. This intuition is expressed with brackets: $[LF; ATT]; WK$. The "retraction" observed in example I still fits this pattern, but the increased difficulty in overriding is only implicit.

Lascarides and Asher (1993) illustrate patterns of defeasible rule interactions. The two inference patterns most relevant here are the Nixon Diamond and the Penguin Principle defined below ($\varphi \to \psi$ means "if φ, then indefeasibly ψ," and $\varphi > \psi$ means "if φ, then normally ψ.").[27]

Nixon Diamond A conflict is unresolved resulting in an ambiguity or incoherence: $(\varphi > \chi) \wedge (\psi > \neg\chi) \supset (\varphi, \psi > \chi \wedge \neg\chi)$.

Penguin Principle A conflict is resolved by the more specific principle defeating the more general one:[28]

$$(\varphi \to \psi) \wedge (\varphi > \chi) \wedge (\psi > \neg\chi) \supset (\varphi, \psi > \chi).$$

On their account, any resolution of a conflict between two defeasible rules should be a case of the Penguin Principle. Does it explain all the conflict resolution patterns observed in pronoun interpretation?

The Penguin Principle explains some of the conflict resolution patterns — for instance, the knowledge about specific agents, Terminator and Arnold, override the generic causal knowledge about hitting (examples G and H). There may also be a remote conceptual connection between the Penguin Principle and the pattern $[LF; ATT]; WK$ in the following line — grammatical preferences (ATT and LF) tend to be more abstract than commonsense preferences (WK) about particular types of eventualities, so the more specific support wins (Kameyama et al., 1993). However, the LF, ATT, and WK rules apply on different

---

[26]$\varphi; \psi[X]$ means $\psi[\varphi[X]]$, where $p[X]$ means $X \cap [[p]]$ (update state X with p).

[27]In these definitions, I use the notations from Asher and Morreau's (1993) Commonsense Entailment (CE) logic as a theoretical meta–formalism without strictly adhering to the CE ontology.

[28]It follows from Cautious Monotonicity [A⇒B, A⇒C / A,B⇒C]:
$(\varphi \to \psi) \wedge (\varphi > \chi) \supset (\varphi, \psi > \chi)$ because $(\varphi \wedge \psi) \leftrightarrow \varphi$.

data structures, and cannot always be reduced to an indefeasible implication ($\varphi \to \psi$) as required in the Penguin Principle. For instance, $hittee(x)$ can be $subject(x)$ or $\neg subject(x)$ depending on the sentence structure, so we cannot say that $hittee(x)$ implies $\neg subject(x)$ to derive the overriding pattern in example F. What additional kinds of conflict resolution inferences do we have then?

There are two additional conflict resolution patterns observed in the present examples, which I will call the *Indefeasible Override* and the *Defeasible Override*, defined below:

Indefeasible Override An indefeasible principle overrides a defeasible one: $(\varphi \to \chi) \wedge (\psi > \neg\chi) \supset (\varphi, \psi \to \chi)$.

Defeasible Override Given an explicit overriding relation, one defeasible principle defeats another (even when $\psi > \neg\chi$): $(\psi; \varphi) \wedge (\varphi > \chi) \supset (\varphi, \psi > \chi)$.

The Indefeasible Override follows from the monotonicity of classical implication ($\varphi \to \chi \supset \varphi, \psi \to \chi$), and is an inherent principle in any nonmonotonic logic. It predicts the fact that the SEM rules override all the defeasible rules (examples C and E). The Defeasible Override captures a certain *a priori* given "ranks" or "priorities" among different sources of information, using the *dynamic override* (;) operator, where $\varphi; \psi$ means "ψ overrides φ." It is motivated by the view that preferences come from different sources, and are associated with different "degrees of defeasibility" not necessarily in terms of the Penguin Principle.[29] It enables us to state the override pattern $[LF; ATT]; WK$ while allowing a varying degree of difficulty for WK's overriding. I hope to define a logical system that axiomatizes these conflict resolution inferences.

7   Further Questions

A number of questions related to the present topic have not been discussed. The first are *logical questions*. What are the connections with *update logics* (e.g., Veltman, 1993)? We can see that the grammar subsystem supports *straight updating*, whereas the pragmatics subsystem supports *preferential updating* or *upgrading* (van Benthem et al., 1993). The preference interaction patterns discussed here can perhaps be formulated as fine–grained upgrading inferences during utterance interpretation within the proposed utterance interpretation architecture. Can my proposal be couched in a system of *preferential dynamic logic* that combines elements of dynamic semantic theories and preferential models (e.g., McCarthy, 1980; Shoham, 1988)? Does the context as a multicomponent data structure proposed here also support the

---

[29]Gärdenfors and Makinson's (1994) use of expectation ordering in preferential reasonning achieves essentially the same effect.

general contextual inferences such as *lifting* in the context logic (e.g., McCarthy, 1993; Buvač and Mason, 1993)?

There are also *computational questions*. Does the proposed discourse processing architecture with explicit contextual control of inferences actually *help manage* the computational complexity of the nonmonotonic reasoning in the pragmatic rule interactions?

Finally, a *cognitive question* — Does the proposed discourse processing architecture naturally extend to a more elaborate many–person discourse model that addresses the issue of coordinating different *private* contexts (e.g., Perrault, 1990; Thomason, 1990; Jaspars, 1994)?

8 Conclusions

A discourse processing architecture with desirable computational properties consists of a grammar subsystem representing the space of possibilities and a pragmatics subsystem representing the space of preferences. Underspecified logical forms proposed in the computational literature define the grammar–pragmatics boundary. Utterance interpretation induces a complex interaction of defeasible rules in the pragmatics subsystem. Upon scrutiny of a set of examples involving intersentential pronominal anaphora, I have identified different groups of defeasible rules that determine the preferred transitions of different components of the dynamic context. There are grammatical preferences inducing fast entity–level inferences only indirectly suggesting the preferred discourse model, and commonsense preferences inducing slow proposition–level inferences directly determining the preferred discourse model. The attentional state in the context supports the formulation of attentional rules that significantly affect pronoun interpretation preferences. The observed patterns of conflict resolution among interacting preferences are predicted by a small set of inference patterns including the one that assumes an explicitly given overriding relation between rules or rule groups. In general, I hope that this paper has made clear some of the *actual* complexities of interacting preferences in linguistic pragmatics, and that the discussion has made them sufficiently sorted out for further logical implementations.[30]

References

Alshawi, Hiyan. 1990. Resolving Quasi Logical Forms. *Computational Linguistics*, 16(3), 133–144.

Alshawi, Hiyan. ed. 1992. *The Core Language Engine*, The MIT Press, Cambridge, MA.

---

[30] In the longer version of this paper (Kameyama, 1994a), a logical implementation of the preferential rule interactions is proposed using prioritized circumscription (McCarthy, 1980, 1986; Lifschitz, 1988), a nonmonotonic reasoning formalism in AI.

Alshawi, Hiyan, and Richard Crouch. 1992. Monotonic Semantic Interpretation. In *Proceedings of the 30th Annual Meeting of the Association for Computational Linguistics*, Newark, DE, 32–39.

Alshawi, Hiyan, and Jan van Eijck. 1989. Logical Forms in the Core Language Engine. In *Proceedings of the 27th Annual Meeting of the Association for Computational Linguistics*, Vancouver, Canada, 25–32.

Asher, Nicholas and Michael Morreau. 1993. Commonsense Entailment: A Modal Theory of Nonmonotonic Reasoning. In *Proceedings of the International Joint Conference on Artificial Intelligence*, Chambray, France, 387–392.

van Benthem, Johan, Jan van Eijck, and Alla Frolova. 1993. Changing Preferences. Report CS-R9310. Institute for Logic, Language and Computation, University of Amsterdam.

Brennan, Susan, Lyn Friedman, and Carl Pollard. 1987. A Centering Approach to Pronouns. In *Proceedings of the 25th Annual Meeting of the Association for Computational Linguistics*, 155–162.

Buvač, Saša and Ian Mason. 1993. Propositional Logic of Context. In *Proceedings of the 11th National Conference on Artificial Intelligence*, 412–419.

Carter, David. 1987. *Interpreting Anaphors in Natural Language Texts*. Ellis Horwood, Chichester, Sussex, UK.

Gärdenfors, Peter and David Makinson. 1994. Nonmonotonic Inference Based on Expectations. *Artificial Intelligence*, 65, 197–245.

Groenendijk, Jeroen and Martin Stokhof. 1991. Dynamic Predicate Logic. *Linguistics and Philosophy*, 14, 39–100.

Grosz, Barbara, Aravind Joshi, and Scott Weinstein. 1983. Providing a Unified Account of Definite Noun Phrases in Discourse. In *Proceedings of the 21st Meeting of the Association of Computational Linguistics*, Cambridge, MA, 44–50.

Grosz, Barbara, Aravind Joshi, and Scott Weinstein. 1986. Towards a computational theory of discourse interpretation. Unpublished manuscript. [The final version to appear in *Computational Linguistics* under the title "Centering: A Framework for Modelling the Local Coherence of Discourse"]

Grosz, Barbara and Candy Sidner. 1986. Attention, Intention, and the Structure of Discourse. *Computational Linguistics*, 12(3), 175–204.

Heim, Irene. 1982. *The Semantics of Definite and Indefinite Noun Phrases*, Ph.D. Thesis, University of Massachusetts, Amherst.

Hobbs, Jerry. 1978. Resolving Pronoun References. *Lingua*, 44, 311–338. Also in B. Grosz, K. Sparck-Jones, and B. Webber, eds., *Readings in Natural Language Processing*, Morgan Kaufmann, Los Altos, CA, 1986, 339–352.

Hobbs, Jerry. 1983. An Improper Treatment of Quantification in Ordinary English. In *Proceedings of the 21st Meeting of the Association of Computational Linguistics*, Cambridge, MA, 57–63.

Hobbs, Jerry, Mark Stickel, Doug Appelt, and Paul Martin. 1993. Interpretation as Abduction. *Artificial Intelligence*, 63, 69–142.

Hudson D'Zmura, Susan. 1988. *The Structure of Discourse and Anaphor Resolution: The Discourse Center and the Roles of Nouns and Pronouns.* Ph.D. Thesis, University of Rochester.

Hwang, Chung Hee, and Lenhart K. Schubert. 1992a. Episodic Logic: A Comprehensive Semantic Representation and Knowledge Representation for Language Understanding. Technical Report, Dept. of Computer Science, University of Rochester, Rochester, NY.

Hwang, Chung Hee, and Lenhart K. Schubert. 1992b. Episodic Logic: A Situational Logic for Natural Language Processing. In P. Aczel, D. Israel, Y. Katagiri, and S. Peters, eds., *Situation Theory and its Applications*, Volume 3, CSLI, Stanford, CA.

Jaspars, Jan. 1994. *Calculi for Constructive Communication*, Ph.D. Thesis, University of Tilberg.

Joshi, Aravind, and Steve Kuhn. 1979. Centered Logic: The Role of Entity Centered Sentence Representation in Natural Language Inferencing. In *Proceedings of International Joint Conference on Artificial Intelligence*, Tokyo, Japan, 435–439.

Joshi, Aravind, and Scott Weinstein. 1981. Control of Inference: Role of Some Aspects of Discoruse Structure — Centering. In *Proceedings of International Joint Conference on Artificial Intelligence*, Vancouver, Canada, 385–387.

Kameyama, Megumi. 1985. *Zero Anaphora: The Case of Japanese.* Ph.D. Thesis, Stanford University.

Kameyama, Megumi. 1986. A Property-sharing Constraints in Centering. In *Proceedings of the 24th Annual Meeting of the Association for Computational Linguistics*, New York, NY, 200–206.

Kameyama, Megumi. 1994a. Indefeasible Semantics and Defeasible Pragmatics. CWI Report CS-R9441 and SRI Technical Note 544.

Kameyama, Megumi. 1994b. Stressed and Unstressed Pronouns: Complementary Preferences. In P. Bosch and R. van der Sandt, eds., *Focus and Natural Language Processing*, Institute for Logic and Linguistics, IBM, Heidelberg, 475-484.

Kameyama, Megumi. 1995. The Syntax and Semantics of the Japanese Language Engine. In R. Mazuka and N. Nagai, eds., *Japanese Sentence Processing*, Lawrence Erlbaum Associates, Hillsdale, NJ, 153–176.

Kameyama, Megumi, Rebecca Passonneau, and Massimo Poesio. Temporal Centering. 1993. In *Proceedings of the 31st Meeting of the Association of Computational Linguistics*, Columbus, OH, 70–77.

Kamp, Hans. 1981. A Theory of Truth and Semantic Representation. In J. Groenendijk, T. Janssen, and M. Stokhof, eds., *Formal Methods in the Study of Language*, Mathematical Center, Amsterdam, 277–322.

Kamp, Hans, and Uwe Reyle. 1993. *From Discourse to Logic*, Kluwer Academic Publishers, Dordrecht.

Kehler, Andrew. 1993. The Effect of Establishing Coherence in Ellipsis and Anaphora Resolution. In *Proceedings of the 31st Meeting of the Association of Computational Linguistics*, Columbus, OH, 62–69.

Lascarides, Alex, and Nicholas Asher. 1993. Temporal Interpretation, Discourse Relations, and Commonsense Entailment. *Linguistics and Philosophy*, 16, 437–493.

Lewis, David. 1979. Scorekeeping in a Language Game. *Journal of Philosophical Logic*, 8, 339–359.

Lifschitz, Vladimir. 1988. Circumscriptive Theories: A Logic–Based Framework for Knowledge Representation. *Journal of Philosophical Logic*, 17(4), 391–442.

Maxwell, John, and Ronald Kaplan. 1993. The Interface between Phrasal and Functional Constraints. *Computational Linguistics*, 19(4), 571–590.

McCarthy, John. 1980. Circumscription–A Form of Non-monotonic Reasoning. *Artificial Intelligence*, 13(1,2), 27–39.

McCarthy, John. 1986. Applications of Circumscription to Formalizing Commonsense Knowledge. *Artificial Intelligence*, 28, 89–116.

McCarthy, John. 1993. Notes on Formalizing Context. In *Proceedings of the International Joint Conference on Artificial Intelligence*, Chambray, France, 555–560.

Pereira, Fernando and Martha Pollack. 1991. Incremental Interpretation. *Artificial Intelligence*, 50, 37–82.

Pereira, Fernando, and David Warren. 1980. Definite Clause Grammars for Language Analysis. *Artificial Intelligence*, 13, 231–278.

Perrault, C. Ray. 1990. An Application of Default Logic to Speech Act Theory. In P. Cohen, J. Morgan, and M. Pollack, eds., *Intentions in Communications*, MIT Press, Cambridge, MA, 161–185.

Poesio, Massimo. 1993. *Discourse Interpretation and the Scope of Operators*. Ph.D. Thesis, University of Rochester.

Pruest, Hub. 1992. *On Discourse Structuring, VP Anaphora and Gapping*. Ph.D. Thesis, University of Amsterdam.

Reyle, Uwe. 1993. Dealing with Ambiguities by Underspecification: Construction, Representation, and Deduction. *Journal of Semantics*, 10(2).

Sag, Ivan, and Jorge Hankamer. 1984. Toward a Theory of Anaphoric Processing. *Linguistics and Philosophy*, 7, 325–345.

Sgall, Petr, Eva Hajičová, and Jarmila Panevová.. 1986. *The Meaning of the Sentence in its Semantic and Pragmatics Aspects*, Reidel, Dordrecht and Academia, Prague.

Shoham, Yoav. 1988. *Reasoning about Change: Time and Causality from the Standpoint of Artificial Intelligence*, MIT Press, Cambridge, MA.

Sidner, Candy. 1983. Focusing in the comprehension of definite anaphora. In M. Brady and R. C. Berwick, eds., *Computational Models of Discourse*, The MIT Press, Cambridge, MA, 267–330.

Stalnaker, Robert, C. 1972. Pragmatics. In Davidson and Harman, eds., *Semantics of Natural Language*, Reidel, Dordrecht, 380–397.

Stalnaker, Robert, C. 1980. Assertion. In P. Cole, ed., *Syntax and Semantics Vol.9: Pragmatics*, Academic Press, New York, 315–332.

Thomason, Richmond. 1990. Propagating Epistemic Coordination through Mutual Defaults I. In R. Parikh, ed., *Theoretical Aspects of Reasoning about Knowledge, Proceedings of the Third Conference (TARK 1990)*, Morgan Kaufmann, Palo Alto, CA, 29–38.

Veltman, Frank. 1993. Defaults in Update Semantics. Manuscript, Department of Philosophy, University of Amsterdam. To appear in the *Journal of Philosophical Logic*.

Wilks, Yorick. 1975. A Preferential, Pattern–seeking Semantics for Natural Language Inference. *Artificial Intelligence*, 6, 53–74.

6

Resumptive Quantifiers in Exception Sentences

FRIEDERIKE MOLTMANN

1 Introduction

The issue whether natural language has true instances of polyadic quantification is a matter of controversy. The literature on polyadic quantification is focused mainly on logical issues, without studying in greater detail actual constructions of polyadic quantification in natural languages.[1]

This paper is a study of what appears to be a clear instance of polyadic quantification in English, namely, polyadic quantification in exception sentences. In connection with this, this paper also presents a semantic analysis of exception constructions in general.[2]

Exception constructions without polyadic quantifiers are exemplified in (1) and in (2):

(1)　　a. every student except John
　　　　b. no student but John and Bill
　　　　c. Except for John and Bill, Mary knows every student.
　　　　d. John came. Nobody else came.

I would like to thank Johan van Benthem, Kit Fine, Makoto Kanazawa, Ed Keenan, Dag Westerståhl, an anonymous reviewer, and audiences of the Workshop on Language, Logic and Information at Stanford University, 1993, and the Conference on Logic and Linguistics at Ohio State University, 1993, for discussion and comments on an earlier version of this paper.

[1]For logically oriented studies of polyadic quantifiers with applications to natural language see in particular van Benthem (1989) and Keenan (1987, 1992). Cases of polyadic quantification in natural language are discussed also in Clark/Keenan (1985/6), Higginbotham/May (1981), May (1989), Nam (1991), and Srivastav (1991).

[2]A more linguistically oriented presentation of this analysis with further empirical applications is given in Moltmann (to appear); see also Chapter 5 of Moltmann (1992).

Quantifiers, Deduction, and Context
Makoto Kanazawa, Christopher Piñón, and Henriëtte de Swart, editors
Copyright © 1996, CSLI Publications

      e. John came. Otherwise, nobody came.

(2)   a. every student but one

       b. every student except at most two

(1a-e) are what I call *simple exception constructions*. Here an *exception phrase*, e.g. *except/but John, except for John and Bill, otherwise*, or *else* associates with a single NP, which denotes a monadic quantifier, and the exception phrase specifies a single entity (or a set of entities) as the exception(s). (2a-b) are somewhat more complex exception constructions, in which the complement of *except* or *but* is a quantified NP.

Exception constructions with polyadic quantifiers are illustrated in (3):

(3)   a. John danced with Mary. Nobody else danced with anybody else.(Keenan 1992)

       b. Every man danced with every woman except John with Mary.

The second sentence of (3a) involves polyadic quantification in that it is equivalent to the claim 'No pair of people danced except for the one consisting of John and Mary', and (3b) involves polyadic quantification in that it is equivalent to the claim 'Every man-woman pair danced except for the one consisting of John and Mary'.

The way I will proceed is as follows. First, I will present a semantic analysis of simple exception constructions and discuss in greater detail the constraints on the quantifier an exception phrase may associate with. Second, I will generalize this analysis to the constructions in (2). Third, I will show how the analysis carries over to exception constructions with polyadic quantifiers as in (3). I will also present various results concerning exception sentences with polyadic quantifiers; in particular, I will show that even though the polyadic quantifiers to which exception phrases may apply are iterations of monadic quantifiers and hence reducible polyadic quantifiers in Keenan's (1987, 1992) sense, the resulting quantifiers are not reducible; that is, they are not definable as iterations of monadic quantifiers.

2   Basic assumptions and basic properties of exception constructions

In the following, I will assume that an NP such as *every student, some student,* or *no student* denotes a *generalized quantifiers* of type $<1>$, that is, a set of sets. More specifically, *every student* denotes the set of sets containing the set of students as a subset, *some student* the set of sets containing at least one student, and *no student* the set of sets containing no student. *Every, some,* and *no* denote *determiners*, that is, functions from sets (as denoted by the N') to generalized quantifiers (of type <1>).

Exception sentences exhibit a number of semantic properties that any theory has to account for. Before presenting my analysis, let me first introduce the three main semantic characteristics of exception constructions.

2.1 The Negative Condition

The *Negative Condition* simply says that the entities which the exception phrase specifies as the exceptions have to be 'exceptions'; that is, when the associated quantifier is positive, the exceptions should not fall under the predicate, and when the associated quantifier is negative, they should fall under the predicate. Thus, the first sentence in (4a) implies that John did not come and the first sentence in (4b) that John did come:

(4) a. #Every student except John came. Perhaps John also came.
     b. #No student except John came. Perhaps John did not come.

2.2 The Condition of Inclusion

The *Condition of Inclusion* concerns (the most common) exception constructions in which the exception phrase associates with an NP of the form D N'. The condition consists in the requirement that the exceptions belong to the restriction of the quantifier denoted by the associated NP, i.e. the denotation of the N'. Thus, (5a) implies that John is a student, as does (5b):[3]

(5) a. Every student except John came.
     b. No student except John came.

2.3 The Quantifier Constraint

The third and most interesting property of exception constructions is the *Quantifier Constraint*. The Quantifier Constraint says that the associated NP of an exception phrase has to denote a quantifier of a

---

[3]There are apparent exceptions to the Condition of Inclusion such as:

(1) Everybody who was invited except Mary's children came to the party.

(1) is true even if not all of Mary's children were invited. In fact, (1) is equivalent to 'everybody except the children of Mary who were invited came to the party'. So it seems as if the exception set, Mary's children, need not be included in, but has to only overlap with the restriction of the associated quantifier.

However, such apparent counterexamples can easily be explained away by the fact that definite plurals may refer to only part of the group satisfying the relevant descriptive content, a fact which is independent of exception constructions. For example, also in (2), *Mary's children* may refer to only part of Mary's children (those that are in the class):

(2) The students in the class are doing well and Mary's children too.

certain kind; basically, it has to denote a universal quantifier or negative universal quantifier (i.e. the internal negation of a universal quantifier). This holds both for *except-* and *but-*phrases, as in (6a), and, though in not as strict a way, for *except for-*phrases, as in (6b):

(6)   a. Every student / All students / No student / #Some students / #Ten students / #Most students / #A lot of students but / except John came to the party.

       b. Except for John, Mary knows every student / no student / ??most students / ??few students / #ten students / #some students.

The Quantifier Constraint should also cover certain other NPs than those whose determiner is *every, all,* or *no.* In particular, it should cover certain coordinated NPs and NPs with wide scope quantified specifiers and complements, as in the acceptable examples among the following:[4]

(7)   a. every man and every woman / #some man and some woman except the parents of John

       b. every president's wife / #some president's wife except Hillary Clinton

       c. the wife of every president / #some president except Hillary Clinton

Coordinated NPs with exception phrases are discussed in more detail in Section 1.3.3., and those of the type in (7b, c) in Moltmann (to appear).

3   Semantic analysis of simple exception constructions

3.1   Exception phrases as operators on generalized quantifiers

There are a number of proposals concerning the semantics of exception constructions in the literature, most notably the analyses by Hoeksema (1987, 1989, 1991) and von Fintel (1993). These analyses are discussed and compared to the present one in Moltmann (1992, to appear). In this paper, I will limit myself to presenting my own account and showing

---

[4] Another piece of evidence that the Quantifier Constraint is not tied to the occurrence of *every, all,* and *no* is that NPs with the cardinality quantifier *the maximal number of* allow for exception phrases, as in (1) (in, let's say, a situation in which every student is assigned a different set of problems and has to try to solve as many of them as possible):

(1)   Every student solved the maximal number of problems except one.

how the three properties of exception phrases mentioned in the previous section are derivable from it.

I will first restrict my attention to simple exception constructions. This means that I can assume that the complement of *except* or *but* denotes a specific set, the *exception set*. Thus, in (1a), *John* is taken to denote the set {John}, and in (1b) and (1c), *John* and *Bill* is taken to denote the set {John, Bill}.

The main idea of the present proposal is that exception phrases are operators on generalized quantifiers, yielding what I will call an *exception quantifier*. More specifically, exception phrases map a generalized quantifier onto another generalized quantifier (an exception quantifier) by doing either one of the following two things:

[1] *subtract* the exceptions from the sets in the generalized quantifier

[2] *add* the exceptions to the sets in the generalized quantifier.

[1] applies when the quantifier is positive as in the case of *every student except John*. [2] applies when the quantifier is negative as in the case of *no student except John*.

In order for either [1] or [2] to apply, however, a certain precondition has to be satisfied. For [1], this precondition is that the exceptions be included in every set in the quantifier; for [2], the precondition is that the exceptions be excluded from every set in the quantifier. I will call the condition of homogeneous inclusion or exclusion the *Homogeneity Condition*, which is formally defined as follows (for nonempty quantifiers):

(8) *Definition*
    For a quantifier Q of type $<1>$, $Q \neq \emptyset$, and a set A, $Hom(Q, C)$ (Q satisfies the *Homogeneity Condition* with respect to C) iff either for every $V \in Q$, $C \subseteq V$ (*homogeneous inclusion*) or for every $V \in Q$, $C \cap V = \emptyset$ (*homogeneous exclusion*).

The Homogeneity Condition is crucial for explaining the Quantifier Constraint and also the Condition of Inclusion. An exception NP not meeting the Homogeneity Condition will have an undefined denotation, and hence any sentence containing it will be neither true nor false, but rather undefined. This corresponds to the fact that speakers generally judge sentences not meeting the Quantifier Constraint or the Condition of Inclusion (such as the marked examples in (6) and (7)) as unacceptable, rather than false.

The denotation of *except* (and *but*) can now be defined as a function mapping a set C to a function from generalized quantifiers to general-

ized quantifiers, as in (9), where Q is the associated quantifier and C
the exception set as denoted by the *except*-complement:

(9) *The denotation of 'except' (first definition)*
  For a set C and a generalized quantifier Q of type <1>,

$$([except](C))(Q) \begin{cases} = \{V \setminus C \mid V \in Q\}, \text{if } C \subseteq V \text{ for all } V \in Q \\ = \{V \cup C \mid V \in Q\}, \text{if } C \cap V = \emptyset \text{ for all } V \in Q \\ = \text{undefined otherwise.}^5 \end{cases}$$

To see how (9) works, consider first (10):

(10) every student except John

Given that John is a student, he will be included in every set in the

---

[5] (9) involves a disjunctive condition on the interpretation of exception phrases,
depending on whether homogeneous inclusion or exclusion obtains. This raises the
question whether there is a way of formulating the semantics of exception phrases
in a uniform manner. Let me suggest such a reformulation. This reformulation
relies on assigning expressions denotation pairs consisting of a positive and a negative extension. This method is often used for the construction of partial models (for instance for the treatment of partiality in the context of perception or
other attitude reports; see, for instance, Muskens (1989), and for partial quantifiers in particular, van Eijck (1991).) A predicate like *come* will be assigned a pair
$< [come]^+, [come]^- >$, where $[come]^+$, the positive extension of *come*, is the set of
entities that come and $[come]^-$, the negative extension of *come*, the set of entities
that do not come. Also an NP such as *every student* will be assigned a pair of
quantifiers $< [everystudent]^+, [every student]^- >$, where $[every student]^+$ consists
of those sets that include every student, and $[every student]^-$ of those sets that
exclude every student. (Hence $[every student]^- = [no student]^+$.) The truth conditions for a simple intransitive sentence then are as in (1), where π_1 is the function
that maps a pair to its first projection and π_2 the function that maps a pair to its
second projection:

(1) $[every \ student \ came] = 1$ iff $\pi_1([came]) \in \pi_1([every \ student])$
  and $\pi_2([came]) \in \pi_2([every \ student])$

An exception phrase now operates on both the positive and the negative extension
of a quantifier. For a quantifier satisfying the Homogeneity Condition, both the
operation of subtraction and the operation of addition will apply to either the
positive or the negative extension. This then allows for the following non-disjunctive
formulation of the denotation of *except*:

(2) *Uniform treatment of 'except'*
  For a set C and a generalized quantifier Q,
  $[except](C)(< Q, Q\neg >)$

$$\begin{cases} = \text{the quantifier pair @ such that for some i} \in \{1,2\} : \text{for all } V \in \pi_i(< Q, \\ Q\neg >), C \subseteq V \text{ and } \pi_i(@) = \{V \setminus C \mid V \in \pi_i(< Q, Q\neg >)\}\}], \text{and for every j} \in \\ \{1,2\}, i \neq j, \pi_j(@) = \pi_i(@)\neg \\ = \text{undefined otherwise} \end{cases}$$

This reformulation (though not terribly elegant) shows that the exception operation
as it is proposed in this paper can be conceived of as a uniform semantic operation.
(Admittedly, though, (2) still contains a hidden disjunction, namely the existential
quantifier ranging over the two projection indices of the quantifier pair.)

quantifier [*every student*]. Hence, John can and will be subtracted from every set in [*every student*].

Now consider (11):

(11) no student except John

Given that John is a student, he will be excluded from every set in the quantifier [*no student*]. Hence, John can and will be added to every set in [*no student*].

We will now see that (9), with some modifications, allows for a derivation of the three properties of exception phrases mentioned above.

3.2 Deriving the three basic properties of exception constructions

3.2.1 The Negative Condition

It is obvious that the Negative Condition follows from (9). A generalized quantifier determines which predicates make the sentence true. If the exceptions have been taken away from all the sets in the generalized quantifier, then, since the predicate extension should be among those sets, the exceptions should not fall under the predicate. If the exceptions have been added to all the sets in the generalized quantifier, then, for the same reason, the exceptions should fall under the predicate.

3.2.2 The Condition of Inclusion

Also the Condition of Inclusion follows from the account given so far; more precisely, it follows from the Homogeneity Condition. It does so in a slightly less trivial way than in the case of the Negative Condition:

(12) *Proposition*
For a conservative determiner D, a set A such that $D(A) \neq \emptyset$, and a set C, if Hom(D(A), C), then $C \subseteq A$.
Proof: First case. Let A and C be sets such that C is homogeneously included in D(A). Since $D(A) \neq \emptyset$, there is a $X \in D(A)$. By Conservativity, $X \cap A \in D(A)$. Given the assumption, $C \subseteq X \cap A$, and hence $C \subseteq A$. *Second case.* Let A and C be sets such that C is homogeneously excluded from D(A). But then C is homogeneously included in the *internal negation* of D (D¬) applied to A, $(D\neg)(A) = \{E \setminus B \mid B \in D(A)\}$ (for E the domain of entities). Since conservativity is preserved by internal negation, the second case reduces to the first.

3.2.3 The Quantifier Constraint

From (9) it is not yet possible to fully derive the Quantifier Constraint. (9) excludes exception constructions not meeting the Quantifier Constraint only for appropriate models. For instance, (9) rules out (13a)

only in a model with more than ten students and (13b) only in a model
with more than two students:

(13)   a. #ten students except John
        b. #most students except John

(13a) is ruled out in a model with more than ten students since in
such a model John (whether he is a student or not) will be included in
some, but not in *all*, sets in the denotation of *ten students*, violating
the Homogeneity Condition. This means that (13a) will not have a
denotation in that model, and hence any sentence containing (13a) will
not have a truth value. (13b) is ruled out in a model with more than
two students since in such a model John will be included in some, but
not in *all*, sets in the denotation of *most students* (again regardless of
whether John is a student or not). Thus, for appropriate models, (9)
explains the unacceptability of (13a) and (13b), and similar examples.

However, the account in (9) fails when such exception NPs are evaluated with respect to other, smaller models. For example, if (13a) is
evaluated with respect to a model which contains exactly ten students
with John among them, John will be included in all sets in the denotation of *ten students*, and hence the Homogeneity Condition will
be satisfied. Similarly, if (13b) is evaluated with respect to a model
with exactly two students and John is one of them, then John will be
included in all sets in the denotation of *most students*, allowing the
Homogeneity Condition to be satisfied.

Thus, (9) has to be strengthened in order to rule out such cases.
The obvious way is by requiring that the Homogeneity Condition should
not only be satisfied with respect to the intended model (thus allowing
it to be 'accidentally' satisfied), but with respect to other models as
well.[6] But which models? One option yields obvious problems, namely

---

[6]One might think of alternative ways of ruling out the unacceptable examples.
For instance, one might argue that the relevant quantifiers are subject to certain
presuppositions which have to be satisfied by the universe in question. *Ten student*
would presuppose that the universe contains more than ten student and *most students* that it contains more than two students. Exception phrases would be defined
only for quantifiers on universes in which the presuppositions of those quantifiers
are satisfied. However, the problem with this proposal is that the speaker might
not know how many students there actually are in the universe, and in such a case,
the presumed presupposition certainly need not be satisfied. If the speaker does not
not know the actual number of students and it turns out that there are only ten
students or only two students, then a sentence such as *ten students took the exam*
or *most students took the exam* intuitively is not undefined, but has a definite truth
value.

Another option, suggested by Ed Keenan (p.c.), is to impose a nonrelational
semantic condition on the associated NP. Such a condition, most plausibly, would
require that the NP accepting an exception phrase denote a filter or an ideal in every
model (cf. Section 3.3.). For example, *every student* always denotes a filter and *no
student* an ideal; but *ten students* and *most students* denote neither a filter nor an

that the Homogeneity Condition should be satisfied in *all* models. In models in which John is not a student, the Homogeneity Condition is not met for *every student but John*. And certainly, John's being a student should not be considered a logical truth. A weaker condition is required, namely, basically, that the Homogeneity Condition hold in the *extensions* of the intended model.[7] For the purpose of stating this condition, it is useful to now conceive of a generalized quantifier as a function(al) mapping a universe M to a set of subsets of M, as in (14):

(14) *Definition*

A *generalized quantifier* Q of type <1> is a functional Q which assigns to every nonempty set M a set Q^M of subsets of M.

(9) is then to be replaced by the following definition, where Q is the associated generalized quantifier and C the exception set:

(15) *The denotation of 'except' (revised definition)*

For a quantifier Q of type <1>, a nonempty set C, and a universe M, $[except](C)(Q^M)$

$$
\begin{cases}
= \{V\backslash C \mid V \in Q^M\} \text{if for every extension M' of M,} \\
\text{for every } V \in Q^{M'}, C \subseteq V \\
= \{V \cup C \mid V \in Q^M\} \text{if for every extension M' of M,} \\
\text{for every } V \in Q^{M'}, C \cap V = \emptyset \\
= \text{undefined otherwise}
\end{cases}
$$

This is the basic analysis of exception constructions. For the examples under consideration it applies rather smoothly. However, as we will see in the next section, there are certain empirical problems with (15) concerning the behavior of exception phrases with conjoined NPs. These problems require yet a further condition on the applicability of the exception operation.

3.3 An additional condition on the application of the exception operation

In order to discuss the new data with conjoined NPs, it is helpful to give an alternative characterization of the quantifiers satisfying the Homogeneity Condition with respect to some exception set. For this purpose, I introduce the following notions:

(16) *Definition*

Let E be a nonempty set and $P(E)$ the power set of E.

---

ideal in the models in question. This kind of condition would be a metalinguistic semantic requirement on an exception NP. As we will see in Section 3.3., however, there are serious empirical problems with this proposal.

[7]In Moltmann (to appear), further restrictions are imposed on these extensions.

(i) For $A \in P(E)$, $F \subseteq P(E)$ is the *filter generated by* A iff: F is the smallest set containing A such that for any $B \in F$ and $C \in F$, $B \cap C \in F$, and for any $B \in F$ and any C, if $B \subseteq C$, then $C \in F$.

$(F = \{B \in P(E) \mid A \subseteq B\})$

(ii) For $A \in P(E)$, $I \subseteq P(E)$ is the *ideal generated by* A iff: I is the smallest set containing A such that for any $B \in I$ and $C \in I$, $B \cup C \in F$ and for any $B \in I$ and any C if $C \subseteq B$, then $C \in I$.

$(I = \{B \in P(E) \mid B \subseteq A\})$

Clearly, the maximal quantifiers satisfying the Homogeneity Condition with respect to some set A are [1] the filter generated by A (for homogeneous inclusion) and [2] the ideal generated by the complement of A, i.e., A' (for homogeneous exclusion). Obviously, any quantifier that is a subset of the filter generated by A will also homogeneously include A, and any quantifier that is a subset of the ideal generated by A' will homogeneously exclude A. Moreover, these are the only quantifiers allowing for homogeneous inclusion or exclusion. Thus, the quantifiers meeting the Homogeneity Condition can be characterized as follows:

(17) A generalized quantifier Q satisfies the Homogeneity Condition with respect to a set A iff either (i) or (ii):

    (i) Q is a subset of the filter generated by A

    (ii) Q is a subset of the ideal generated by A'.

We can now turn to the new data that present problems for the Homogeneity Condition or equivalently the characterization in (17) as a precondition on the acceptance of an exception phrase. As (17) makes clear, the set of quantifiers satisfying the Homogeneity Condition (with respect to some set A) is closed under subsets and hence closed under conjunction with other quantifiers. Unfortunately, the data with conjunction do not confirm with this prediction and hence require a strengthening of the Homogeneity Condition. Let us look at the various cases of exception constructions with conjunction.

In the following cases, the account in (15) still makes the right prediction:

(18)   a. every man and every woman except John and Mary

        b. no man and no woman except John and Mary

The denotation of *every man* and *every woman* is the set of sets containing the set of men and the set of women as subsets. *John and Mary* is taken to denote the set {John, Mary}, which is contained in every set in the denotation of *every man and every woman*; hence subtraction can apply, taking away John and Mary from each of these sets, as in (19):

(19) [every man and every woman except John and Mary] = {V \
{John, Mary} | V ∈ [every man] & V ∈ [every woman]}

Similarly, the denotation of *no man and no woman* consists in the set
of sets containing no man and no woman; hence the set {John, Mary}
is excluded from every set in that denotation, and therefore can be
added to every such set.

The account in (15) predicts correctly that with an exception set
with members in the domain of each conjunct quantifier, both conjoined
NP have to denote universal or negative universal quantifiers. In (20),
for example, the Homogeneity Condition is violated (in some relevant
model):

(20) #every man and some woman except John and Mary

In a model with more than one woman, [every man and some woman]
will contain some set containing John, but not Mary; hence no homogeneous inclusion or exclusion.

The account in (15) makes a potentially problematic prediction
about coordinated proper names. If a proper name denotes a generalized quantifier (namely the set of sets containing the relevant individual), then a conjunction of proper names, as in (21), allows for
homogeneous inclusion (in the relevant models) and hence should allow for exception phrases:

(21) #John, Bill, and Max except John

But (21) sounds very odd - as does (22):

(22) #John except John

However, (21) and (22) are very plausibly ruled out by a pragmatic
requirement: an exception phrase should not explicitly specify entities
as exceptions that have already explicitly been mentioned as 'nonex-
ceptions'. Thus, (23) sounds better than (21) or (22):

(23) ?John, Bill, Max, Sue, and Ann, except the oldest

The fact that (23) is still degraded may be attributed to the same
pragmatic requirement being still, in some way, violated.

The same observations can be made for negative disjunctions of
proper names with *neither... nor*, which allow for homogeneous exclusion:

(24) a. #?neither John, Bill nor Max except Bill
      b. ?neither John, Bill, Max, Sue, nor Ann except the oldest.

More serious problems for the account arise when the exception set
does not contain any element in the domain of some of the conjunct
quantifiers. Consider the following data:

(25) a. #every man and some woman except John

    b. #Except for John, Mary saw every man and some woman.

    c. #every man, but not every woman except John

    d. #every man and almost every woman except John

(26) a. every man and every woman except John

    b. no man and no woman except John

    c. every mathematician, every physicist, every theologian, and every biologist except Heisenberg

(27) a. #every man and no woman except John

    b. #no man and every woman except John

The account given so far makes the wrong prediction in the examples in (25) as well as (27). *Every man and some woman* in (25a) denotes a subset of [*every man*] and thus homogeneously includes {John}; hence the unacceptability of (25a) is unexpected. (25b) with a free exception phrase shows that the reason for the unacceptability of (25a) cannot be the formal adjacency between *John* and *some woman*. Similarly for (25c) and (25d). The Homogeneity Condition is too weak to rule out examples such as (25a), (25c), and (25d).

As (26a) and (26b) show, *every man and every woman* and *no man and no woman* do, surprisingly, accept simple exception phrases which relate to only one conjunct. (26c) indicates that this is possible without limit as to the number of conjuncts and the distance of the exception phrase from the conjunct it relates to.

The data in (27), again, show that the Homogeneity Condition is too weak. The set {John} is homogeneously included in the intersection of [*every man*] with [*no woman*] and homogeneously excluded from the intersection of [*no man*] with [*every woman*]. Hence (27a) and (27b) should be acceptable, contrary to the facts.

Thus, for the acceptability of an exception phrase, it is not sufficient that the associated quantifier satisfies the Homogeneity Condition (in the relevant models). The data in (25) and (27) show that an additional condition must obtain, a condition which basically has the effect that the conjunct quantifiers must all be either universal or negative universal. It is not a trivial matter, though, to say what exactly this condition consists in. Let us consider some plausible proposals.

First, one might propose that the condition consists in a restriction on the extension of NPs taking exception phrases. I will call such a condition an *extensional condition*. The most plausible extensional condition is that NPs that accept exception phrases must denote a filter or an ideal (in every model): *every man* denotes a filter and *no man* an ideal, whereas *every man and some woman* and *every man and no woman* denote neither a filter nor an ideal (in every model). Hence the acceptability of the former NPs with exception phrases, but not the latter.

The problem with such an extensional condition, however, is that it implies that conjunctions of NPs denoting universal quantifiers with proper names or other singular terms, as in (28a) and (28b), should accept exception phrases; but the results are equally bad:

(28)   a. #every man and Mary except John
        b. #Except for John, Sue met every man and Mary.

Every student and Mary in (28a) denotes a filter, the filter generated by the set $\{x \mid man(x) \vee x = Mary\}$. (28b), again, shows that the reason for the unacceptability of (28a) could not simply be the formal adjacency between *except John* and *Mary*. No extensional condition, in fact, will be able differentiate between the examples in (26) and those in (28).

The second proposal would be more formal in nature and tie the condition in question to the syntactic form of a coordinated NP. On such a proposal, it would, in some form or other, be required that exception phrases modifying coordinated NPs have to potentially satisfy the Homogeneity Condition with respect to each of the conjuncts. I will call a condition of this sort an *across-the-board condition*. This effect of an across-the-board condition (given it can be stated properly in the first place) would be that each conjunct must denote a universal or a negative universal quantifier. Such a condition would not yet rule out the examples in (26), where each conjunct individually could meet the Homogeneity Condition with respect to some exception set. For such cases, it would have to be required that the Homogeneity Condition be satisfied in the same way by each conjunct.

Even with such a modification, there are other serious problems for an across-the-board condition, namely examples such as (29):

(29) #Except for John, Bill knows everybody who is a linguistics student or Mary.

(29) sounds as bad as (28a) and (28b); but it could not possibly be ruled out by an across-the-board condition. The reason why (29) is bad seems to be the same as for (28a) and (28b); thus, a condition of the across-the-board kind would not be general enough.

What is governing the acceptability of exception phrases in (25)-(29) appears to be a condition on the *intension*, rather than the extension or the form of the NP modified by an exception phrase. That is, exception phrases can apply only to NPs whose intension meets a certain condition. Let us adopt a possible-worlds account of intensions. Then the intension of an NP to which an exception phrase may apply is an *intensional quantifier*, that is, a function mapping a possible world w to an (extensional) generalized quantifier ranging over entities in w.

Given this, the condition in question - for now - seems to be as follows: the intension of an NP accepting an exception phrase must be identical to an intensional quantifier of the form D(P), where D is an (intensional) determiner and P is a *purely qualitative property*. A purely qualitative property is a property which does not essentially make reference to any particular individuals or, in linguistic terms, is a property which can be expressed without reference to any individuals. For example, the property of being a student is a purely qualitative property, but the property of being a student of Mary is not since it essentially involves Mary.

The notion of a purely qualitative property has been made precise by Fine (1977): a purely qualitative property is a property which stays the same under any automorphism (i.e., a permutation of individuals which respects the structure of each world) or, given the linguistic characterization, a property which can be expressed by a formula that contains no names.

As it turns out, however, restricting exception phrases to quantifiers with a purely qualitative property as their restriction is too strong a condition. Sometimes, proper names may be used (in an essential way) in the formulation of the quantifier restriction, as in (30a) and (30b):

(30)   a. John knows everybody at UCLA except Tim.
        b. Mary talked to everybody who knows Bill except Max.

Here the properties corresponding to the quantifier restriction are 'person at UCLA' and 'person who knows Bill', which are both not purely qualitative. What distinguishes the examples in (30) from (29)? The difference lies in that in (30), unlike in (29), the proper names do not denote elements in the quantifier restriction, but only elements that serve to identify the quantifier restriction.

So then we get the following weaker condition on properties that may make up the restriction of an exception quantifier: only those properties are acceptable that do not essentially involve entities that belong to their extension in the actual world. Let us call such properties (relative to a possible world) simply *qualitative properties*.

This notion of a qualitative property, again, can be made precise formally: a qualitative property P in a possible world w is a property which stays the same under any automorphism that only permutes individuals in the extension of P in w (i.e., elements of $P(w)$).

I will call the condition that a quantifier have a restriction corresponding to a qualitative property the *Quality Condition*. So we have:

(31) *Definition*
    For an intensional quantifier Q and a possible world w, *Qual*(Q, w) (Q satisfies the *Quality Condition* with respect to w) iff Q

$= D(P)$ for D an (intensional) determiner and P a qualitative property in w.

Let us apply the Quality Condition to the examples in (25)–(29). *Every man and some woman* in (25a) and *every man and no woman* in (27a) clearly do not express an intensional quantifier of the form $D(P)$. Also *every man and Mary* in (28a), even though it is logically of the form 'every x such that man(x) or x = Mary', does not express an intensional quantifier of the appropriate kind, since 'being a man or being Mary' is not a qualitative property. (29) is ruled out in the same way. By contrast, *every man and every woman* and *no man and no woman* in (26a) and (26b) do express intensional quantifiers satisfying the Quality Condition: their intension is the same as that of *every man or woman* and *no man or woman* respectively, and 'being a man or woman' is a qualitative property.[8]

We see that what is crucial for the acceptability of an exception phrase is not that the associated NP is of the form '*every* N'' or '*no* N'', but rather that it is intensionally identical to a universal or negative universal intensional quantifier satisfying the Quality Condition - in whatever way such an intensional quantifier may be expressed. Note that the Quality Condition alone does not suffice; the restriction to universal or negative universal quantifiers still must be derived from the Homogeneity Condition.

So the precondition on the applicability of the exception operation now is as follows:

(32) *Condition on the applicability of the exception operation (strengthening)*
For a set A, an intensional quantifier Q (of type <1>), a universe M, and a possible world w, $([except](A))(Q^M(w))$ is defined iff for every extension M' of M, $\text{Hom}(A, Q^{M'}(w))$ and $\text{Qual}(Q^M, w)$.

I will now turn to two generalizations of the semantic analysis of simple exception constructions that I have given, namely [1] in order to account for exception constructions in which the complement of *except* is a quantified NP and [2] in order to account for exception constructions in which the associated quantifier is polyadic. When formulating these generalizations, I will, for the sake of simplicity, disregard the

---

[8]There are still potentially problematic cases for the Quality Condition. These are quantifiers whose restrictions is formulated by means of a definite description as in the examples below, which seem as bad as the examples in (28) and (29):

(1) a. #every student and the dean except John
  b. #Except for John, Sue met every student and the dean.
(2) #Except for John, Bill knows everybody who is a student or the dean.

If *the dean* is used attributively (rather than referentially), then 'being a student or the dean' should be a qualitative property. So there may be a puzzle.

Quality Condition and stick to (15) as the point of departure for the generalizations. But (32), appropriately generalized, ultimately must apply to the other cases as well.

4 Generalizing the analysis 1: Exception sentences with quantified *except*-complements

Besides definite NPs, the complement of *except* or *but* may be a quantified NP of a certain type, as in (2a, b) repeated here as (33a, b):[9]

(33)   a. every student but one
       b. every student except at most two

The way the data in (33) will be analysed is by applying the denotation of *except* or *but* not to a set, but rather to a generalized quantifier, in this case, the generalized quantifiers denoted by *one (student)* and by *at most two (students)*. This analysis does not abandon the analysis of simple exception constructions in (15), but rather generalizes it.

The way this is achieved is by letting the exception operation apply *pointwise* to the elements in a particular set of sets obtained from a generalized quantifier. For example, in the case of (33a), given a model in which John and Bill are the only students, the exception operation applies to the set {{John}, {Bill}}. The denotation of *every student but one* is obtained by first subtracting John from every set in [*every student*], yielding a set X, and subtracting Bill from every set in [*every student*], yielding a set Y. Applying set union to X and Y yields the denotation of *every student but one*, and in effect renders this NP synonymous to 'every student but John or every student but Bill' in the given universe, a desired result.

The set to which the exception operation now applies can be obtained from any generalized quantifier by means of an operation W, which is defined in terms of the notion of witness set, the latter itself being defined on the basis of the following notion of live-on set:

(34) *Definition (Barwise/Cooper 1981)*
     A generalized quantifier Q of type $<1>$ *lives on* a set A iff for every X: $X \in Q$ iff $X \cap A \in Q$.

A live-on set for *one student* and *at most two students* is any set containing the set of students.

---

[9] *Except* may also take disjoint complements as in (1):

   (1) every boy except John or Bill

It is tempting to subsume disjoint *except*-complements under the general account of quantified *except*-complements given in this section. However, as I argue in Moltmann (to appear), disjoint *except*-complements behave differently from quantified *except*-complements and require separate treatment.

A witness set is defined as follows (in slight deviation from Barwise/Cooper 1981):

(35) *Definition*
A set W is a *witness set* for a generalized quantifier Q of type <1> iff W $\in$ Q and W $\subseteq$ A, where A is the smallest live-on set of Q.

Thus, the witness sets for *one student* in the relevant universe are {John} and {Bill}, and the witness sets for at *most two students* the sets $\emptyset$, {John}, {Bill}, and {John, Bill}.

The operation W is defined as follows:

(36) *Definition*
For a generalized quantifier Q of type <1>,
W(Q) = {X | X is a witness set for Q}.

The application of pointwise subtraction and addition will for (32a) yield the following denotation:

(37) $[every\ student\ but\ one\ (student)]^M = \cup$ {V \ V' | V $\in$ [*no student*]^M}
V' $\in$ W([*(one) student*]^M)

This denotation is obtained by generalizing the denotation of *except* (and *but*) as in (15) so that [*except*] now maps a generalized quantifier onto a function from generalized quantifiers to generalized quantifiers:[10]

(38) *The denotation of 'except' with quantified except-complements*
For generalized quantifiers Q and Q' of type <1> and a universe M, [*except*]$(Q'^M)(Q^M)$

---

[10]There are apparent exceptions to the Homogeneity Condition as applied to quantified *except*-complements:

(1) Every student except two girls did the homework.

Not every witness set (any set of two girls) is included in the set of students, but only those consisting of girls that are students. In fact, (1) is equivalent to 'every student except two girls that are students'.

However, (1) does not pose a serious threat to the Homogeneity Condition, but rather can be considered a case of local accommodation of the presupposition of *except*. By accommodation, the quantification domain of *two girls* will be restricted to girls that are students. Like definite NPs (cf. Fn.3), quantified NPs may also in other contexts range over only part of the entities satisfying the relevant descriptive content. Cases in point are (2a) (which implies that the two teenagers are excellent students) and (2b) (which, on one reading, only implies that the speaker did not talk to any geniuses that are students):

(2)   a. Today, I talked to a number of excellent students, including two teenagers.
     b. Today, I talked to many students, but no geniuses.

$$\begin{cases} = \cup\{V \setminus V' | V \in Q^M\} \text{if for every extension M' of M,} \\ \text{for } V' \in W(Q'^M) \text{every } V' \in W(Q'^{M'}) \text{and } V' \in Q^{M'}, \\ V' \subseteq V \\ = \cup\{V \cup V' | V \in Q^M\} \text{if for every extension M' of M,} \\ \text{for } V' \in W(Q'^M) \text{every } V' \in W(Q'^{M'}) \text{and } V \in Q^{M'}, \\ V' \cap V = \emptyset \\ = \text{undefined otherwise.} \end{cases}$$

Of course, (38) subsumes the simple exception constructions discussed in Section 2 and 3. In the case of the exception NP *every student except John, John* will now denote the quantifier $\{V \mid j \in V\}$. The application of W to this set yields the set $\{\{j\}\}$. For the exception operation to apply, (38) requires that $\{j\}$ be included in every set in [*every student*], and hence, by (12), in the set [*student*]. The denotation of the exception NP then will be $\{V \setminus \{j\} \mid V \in [every\ student]\}$.[11]

---

[11] Unlike what one would expect from (38), not every quantifier may be an argument of *except* or *but*. There are two kinds of restrictions. The first is what I call the *Minority Requirement* (cf. Moltmann, to appear), which says that the exception sets to a quantifier Q must constitute a minority among the entities in the restriction of Q. The Minority Requirement accounts for the contrast between (1a) and (1b) as well as for the one between (2a) and (2b):

(1)   a.   every boy except at most two
       b.   #every boy except at least two
(2)   a.   ?all twenty of the boys boys except two
       b.   #all twenty of the boys except ten

(1a) is fine as long as two boys are a minority among the boys. (1b) is bad since *at least two* allows the entire set of boys to be the exception set, violating the Minority Requirement. The contrast between (2a) and (2b) shows that the Minority Requirement is independent of the type of quantifier, but is sensitive only to the number of relevant entities.

The second restriction, which is independent of the first, may be described as following: only those quantifiers can be the argument of *except* that can act as dynamic existential quantifiers, that is, quantifiers that support E-type pronouns. These quantifiers include *a, two, at most two, few, exactly two, one or two*, but not *less than three, every*, or *no*. The examples in (3) illustrate the cooccurrence with *except*, and those in (4) the support of E-type pronouns:

(3)   a.   every boy except two / at most two / (only) few / exactly two / one or two
       b.   #every boy except less than three / every ten year old one / no ten year old one
(4)   a.   Two boys / At most two boys / Few boys / Exactly two boys / One or two boys came. *They* sat down.
       b.   If a girl sees two boys / at most two boys / exactly two boys / one or two boys, she admires *them*.
       c.   #Less than three boys came. *They* sat down.
       d.   #If a girl sees less than three boys / every boy, she admires *them*.

The reason for this restriction on the quantifiers is not clear to me. The analysis as it stands predicts that all quantifiers should behave alike, subject perhaps only

The analysis given in (15) will be generalized in yet another direction in the next section, namely to exception constructions in which the exception phrase applies to a polyadic quantifier. This first requires a general discussion of such exception constructions.

5 Generalizing the analysis 2: Exception sentences with polyadic quantifiers

5.1 The data

There is a rather broad range of exception constructions involving polyadic quantification. First, there is the *multiple 'else'-construction* in (39a-c), first noted by Keenan (1992) with (39a):

(39)   a. John danced with Mary. Nobody else danced with anybody else.

        b. John danced with Mary and Bill danced with Sue. Nobody else danced with anybody else.

        c. John did not dance with Mary. Everybody else danced with everybody else.

On one reading, the second sentence of (39a) is equivalent to 'Nobody danced with anybody except for the pair consisting of John and Mary'. On this reading, the two occurrences of *else* act as a single exception phrase, specifying the pair consisting of John and Mary as the exception. This exception phrase applies to a *polyadic quantifier*, namely the universal dyadic quantifier ranging over man-woman pairs. I will take this quantifier to be the denotation of the (discontinuous) sequence consisting of the NPs *every man* and *every woman*, that is, [<*every man, every woman*>]. It is the set whose elements are all binary relations containing the product [*man*] x [*woman*] as a subset.

A dyadic quantifier (a set of two-place relations) is a generalized quantifier of type <2>; a triadic quantifier (a set of three-place relations) is a generalized quantifier of type <3>, and so on.

Polyadic readings are available also with the exception expression *otherwise*:

(40)   a. John danced with Mary. Otherwise, nobody danced with anybody.

        b. John did not dance with Mary. Otherwise, everybody danced with everybody.

        c. John danced with Mary. (?)Otherwise, every man danced with no woman.

---

to seemingly pragmatic requirements such as the Minority Requirement. So there is an issue to be investigated.

In another construction in which exception phrases apply to polyadic quantifiers, *except* is followed by a construction which looks like Gapping:

(41) a. Every man danced with every woman except John with Mary.
  b. No man danced with any woman except John with Mary.
  c. Every man danced with every woman every evening except John with Mary yesterday.
  d. Every man danced with every woman except John with Mary and Bill with Sue.
  e. (?)Every man danced with no woman except John with Mary.

Crucially, these sentences are not equivalent to sentences in which simple exception phrases associate with single NPs, namely to (42a), (42b), (42c), and (42d) respectively:

(42) a. Every man except John danced with every woman except Mary.
  b. No man except John danced with any woman except Mary.
  c. Every man except John danced with every woman except Mary every evening except yesterday.
  d. Every man except John and Bill danced with every woman except Mary and Sue.
  e. Every man except John danced with no woman except Mary.

(41a) and (42a), for example, differ in truth conditions in the following ways: (41a) implies that John did not dance with Mary, whereas (42a) has no such implication; moreover, (41a) implies that John danced with every woman other than Mary, whereas (42a) implies that John either did not dance with every woman other than Mary or did dance with Mary.

Polyadic quantification with exception phrases arguably is also involved in the following cases, examples of the sort noted by Hoeksema (1989) and discussed in greater detail in Moltmann (to appear):

(43) a. No man talked to any woman except Mary.
  b. No man talked about any book to any woman except Mary.
  c. No man talked about any book to any woman except about the bible to Mary.

In these cases, a simple exception phrase seems to apply to a single NP. However, the Quantifier Constraint is not locally satisfied in (43a-c). It is satisfied only by the quantifier denoted by a sequence of more than one NP. In (43a), this is the quantifier denoted by <*no man, any woman*>; in (43b), it is the quantifier denoted by <*no man, any woman*>; and in (43c), it is the quantifier denoted by <*no man, any book, any woman>.* In fact, (43a), (43b), and (43c) are equivalent to (44a), (44b), and (44c) respectively, which are clearly polyadic in form:

(44)   a. No man talked to any woman except one man (or more) to Mary.

      b. No man talked about any book to any woman except one man (or more) about one book (or more) to Mary.

      c. No man talked about any book to any woman except one man (or more) about the bible to Mary.

5.2   Generalizing the exception operation to polyadic quantifiers

I now turn to the formal analysis of exception constructions with polyadic quantifiers. First, I will assume that in examples such as (40a) *every man* and *every woman* are evaluated together as the sequence *<every man, every woman>*, which denotes the dyadic quantifier given in (44):[12]

(45) $[<every\ man,\ every\ woman>] = \{R \mid \{x \mid \{y \mid <x, y> \in R\} \in$
     $[every\ woman]\} \in [every\ man]\}$

Such a denotation can be obtained systematically by applying the iteration operation ·, defined in (46), to the monadic quantifiers denoted by the individual NPs:

(46) *Definition*
     For quantifiers Q_1 and Q_2 or type $<1>$,
     $Q_1 \cdot Q_2 = \{R \mid \{x \mid \{y \mid <x, y> \in R\} \in Q_2\} \in Q_1\}$
   · is associative. That is, $(Q_1 \cdot Q_2) \cdot Q_3 = Q_1 \cdot (Q_2 \cdot Q_3)$.
     The denotation of a sequence of NPs can then be defined as:

(47) $[< NP_1, NP_2 >] = [NP_1] \cdot [NP_2]$

With polyadic quantifiers, the exception phrase (in the simple cases) specifies a set of n-tuples, a relation, as the exception set. The exception operation applies to such a set and a polyadic quantifier in exactly the same way as in the monadic case. The only difference is that homogeneous inclusion or exclusion now holds between a set of relations and a relation and that the exception operation either subtracts a relation from the relations in a set or adds a relation to the relations in a set.

I will assume that what follows *except* in (41a) denotes the relation in (48a). Schematically, the denotation of (41a) then is as in (48b), which is exactly parallel to the monadic cases:

(48)   a. $[<John,\ with\ Mary>] = \{<John,\ Mary>\}$

---

[12]The question of how the denotation of such a sequence of NPs as a polyadic quantifier is compatible with compositionality is discussed and suggestively answered in Moltmann (to appear).

b. ($[danced]$) $\in$ $[except]$ ({<John, Mary>}) ([<every man, every woman>]))

The generalized definition of the denotation of *except* that applies in (48b) is given in (49), which is simply a generalization of (15) from sets and generalized quantifiers of type <1> to relations and generalized quantifiers of type <n>:

(49) *The denotation of 'except' with polyadic associated quantifiers*
For an n-place relation R', a generalized quantifier Q of type <n>, and a universe M,

$$[except](R')(Q^M) \begin{cases} = \{R \setminus R' | R \in Q^M\} \text{ if for every extension} \\ \quad M' \text{of } M, \text{ for every } R \in Q^{M'}, R' \subseteq R, \\ = \{R \cup R' | R \in Q^M\} \text{ if for every extension} \\ \quad M'' \text{of } M, \text{ for every } R' \in Q^{M'}, R \cap R = \emptyset \\ = \text{undefined otherwise.} \end{cases}$$

Does (49) again have to be generalized in a way parallel to the generalization from (15) to (38) for disjoined and quantified complements of *except*? In the next section, we will see that, in fact, there are exception constructions with polyadic quantifiers that require such a generalization.

5.3 Exception phrases with polyadic quantifiers and quantified except-complements

In the exception constructions with apparent Gapping, the material following *except* need not specify a particular set of n-tuples as the exceptions as with definite plurals; it may alternatively consist of a sequence of quantified phrases:

(50)   a. Every man danced with every woman except one professor with one student.
        b. Every man danced with every woman except at most one professor with at most one student.
        c. Every man danced with every woman except one professor with one student or one visitor with one secretary.

The generalization to quantified *except*-complements with polyadic quantifiers is straightforward. The *except*-complement now denotes a polyadic generalized quantifier, as in (51) for (50a):

(51) $[one\ professor\ with\ one\ student]$ = {R | {x | {y | <x, y> $\in$ R} $\in$ $[one\ student]$} $\in$ $[one\ professor]$}

Parallel to (38), the exception operation then has to be redefined as follows:

(52) *The denotation of 'except' for polyadic quantifiers and quantified 'except'-complements*

For generalized quantifiers Q and Q′ of type <n> and a universe M, $[except](Q'^{M})(Q^{M})$

$$
\begin{cases}
= \cup\{R \setminus R'|R \in Q^{M}\}\text{if for every extension M' of M, for R'} \\
\in W(Q'^{M})\text{every } R \in Q^{M'}\text{and } R \in W(Q'^{M}), R' \subseteq R'' \\
= \cup\{R \cup R'|R \in Q^{M}\}\text{if for every extension M' of M, for R'} \\
\in W(Q'^{M})\text{every } R \in Q^{M'}\text{and } R \in W(Q'^{M}), R' \cap R'' = \emptyset \\
= \text{undefined otherwise.}
\end{cases}
$$

(52) constitutes the most general definition of the denotation of *except*.

5.4 On the satisfaction of the 'Quantifier Constraint' by a polyadic quantifier

As in the monadic case, it is a precondition for the exception operation to apply to a polyadic quantifier that the Homogeneity Condition be satisfied. However, given that the polyadic quantifiers in question are not denoted by a single NP, but rather defined in terms of the monadic quantifiers denoted by the NPs in a sequence of several NPs, one may ask the following question: is it possible to predict, given the properties of the monadic quantifiers and their ordering in the sequence, whether the resulting quantifier allows for an exception phrase? Given the joint effect of the Homogeneity Condition and the Quality Condition (as in (32)), this question can in fact be answered. The answer crucially involves the notion of *resumptive quantifier*.

A resumptive quantifier is a polyadic quantifier which can be 'looked upon' as a monadic quantifier ranging over n-tuples. For example, the quantifier denoted by <*every man, every woman*> can be conceived of as a monadic quantifier, namely the universal quantifier ranging over man-woman pairs, i.e., the quantifier EVERY([*man*] x [*woman*]).

A formal definition of resumptive quantifier has been given by Westerståhl (1992). According to this definition, an n-ary polyadic quantifier Q is a resumptive quantifier iff there is an ordinary monadic quantifier Q′ defined on the n-ary Cartesian product of the universe such that Q holds of a relation R just in case Q′ holds of R, where R, in the former case, is 'looked upon' as a relation, but in the latter case, as a set of n-tuples.

This definition of resumptive quantifier requires the conception of generalized quantifier as a functional given earlier in (14), now, of course, generalized to quantifiers of type <n> in the obvious way:

(53)   a. *Definition (Westerståhl 1992)*

A generalized quantifier Q of type <n> (n > 0) is the n-ary *resumption* of a generalized quantifier Q′ of type <1> iff

$Q'^M(R) \leftrightarrow Q^M(R)$ for a universe M and an n-place relation R.

b. *Definition*

A generalized quantifier Q of type $<n>$ $(n > 0)$ is *resumptive* iff there is a generalized quantifier Q' of type $<1>$ such that Q is the n-ary resumption of Q'.

The notion of resumptive quantifier provides a sufficient condition on when a polyadic quantifier accepts an exception phrase: if a polyadic quantifier Q is the resumption of a monadic quantifier Q', then Q accepts an exception phrase whenever Q' is universal or negative universal. Applying this criterion, we can see why the sequences $<every$ *man, every woman>*, $<no$ *man, any woman>*, and $<every$ *man, no woman>* accept exception phrases: they all are resumptions of (negative) universal quantifiers:

(54)   a. $[<every \; man, \; every \; woman>] = \text{EVERY}([man] \times [woman])$
      b. $[<no \; man, \; any \; woman>] = \text{NO}([man] \times [woman])$
      c. $[<every \; man, \; no \; woman>] = (\text{EVERY} \; \neg)([man] \times [woman])$

Thus, $[<every \; man, \; every \; woman>]$ is the resumption of the monadic quantifier EVERY with the restriction $[man] \times [woman]$ applied to the product universe, and so on. Being the resumption of a monadic universal or negative universal quantifier, then, is a sufficient condition for whether a quantifier iteration allows for an exception phrase.

However, the notion of resumptive quantifier not only provides a sufficient condition for whether a quantifier iteration allows for an exception phrase. It also allows for stating a necessary condition. Given (32), a quantifier accepts an exception phrase if and only if it is a (negative) universal quantifier ranging over instances of a qualitative property. In the polyadic case, this means that the quantifier should be a (negative) universal quantifier ranging over n-tuples that satisfy an n-place qualitative property. But this means that it is a necessary (and sufficient) condition for such a polyadic quantifier to be the resumption of a (negative) universal quantifier (with an appropriate restriction). The only way for a sequence of NPs to express such a quantifier is when each NP in the sequence denotes a universal quantifier - with two exceptions: the last NP in the sequence may denote the internal negation of a universal quantifier, and two adjacent NPs in the sequence (NP_k and NP_{k+1}) may be such that NP_k denotes the internal negation of a universal quantifier and NP_{k+1} the external negation of a universal quantifier. So the only quantifier iterations that constitute (negative) universal polyadic quantifiers and hence allow for exception phrases have one of the following forms:[13]

---

[13]The identities in (55b) are due to the following Negation Lemma:

(55)   a. $<$EVERY(A_1), ... , EVERY(A_n)$>$ = (EVERY(A_1), ... ,
          NO(A_k), $\neg$ SOME(A_{k+1}), ... ,EVERY(A_n)$>$

      b. $<$EVERY(A_1), ... , EVERY(A_n)$>$ $\neg$ = $<$EVERY(A_1),
          ... , EVERY(A_{n-1}), NO(A_n)$>$ = $\neg$ $<$SOME(A_1), ... ,
          SOME(A_n)$>$ = $<$ $\neg$ SOME(A_1), ... , SOME(A_n)$>$ =
          $<$NO(A_1), ANY(A_2), ... , ANY(A_n)$>$

The following data illustrate somewhat further the restrictions on
quantifier iterations accepting exception phrases:

(56)   a. ??Every man danced with at most two women except John
          with Mary, Sue, and Claire.

      b. ??Every man danced with at least two women except John
          with only Mary.

      c. ??No man danced with at most two women except John with
          exactly one.

      d. ??No man danced with at least two women except John with
          Sue, Mary, and Claire.

None of the quantifier iterations in these examples are resumptive (negative) universal quantifiers. So the criterion introduced above explains
why they do not accept exception phrases.

Exception phrases are excluded with these quantifiers iterations also
by the Homogeneity Condition. Consider the quantifier Q denoted by
$<$*no man, at most two women*$>$ in (56a). A relation R is in Q just in
case R does not contain three pairs $<$m, w$>$, $<$m, w'$>$, and $<$m, w''$>$,
where m is a man and w, w', and w'' are distinct women. This means
that the exception relation \{$<$John, Mary$>$, $<$John, Sue$>$, $<$John,
Claire$>$\} cannot be included in every relation in Q; but it also means
that it cannot be excluded from every such relation. Q will, for example,
include a relation R containing $<$John, Mary$>$, but not $<$John, Sue$>$
and some other relation R' containing $<$John, Sue$>$, but not $<$John,
Mary$>$. Thus, the Homogeneity Condition is not satisfied. Similar
considerations show that the Homogeneity Condition is not satisfied
with any of the other examples in (56).

A slight problem with the examples in (56) is that they are not quite
as bad as predicted. But the reason why they seem to be marginally
possible may be the following: in these examples, only *except John* acts
properly as the exception phrase, relating to *every man* as a monadic
quantifier. The function of *with Mary, Sue, and Claire* and the other
with-phrases may be simply to explicate in which way John is an excep-

---

(1) *Negation Lemma (Westerståhl 1992)*
    For quantifiers $Q_1, ... , Q_n$ of type $<1>$,
      (i) $\neg (Q_1 \cdot ... \cdot Q_n) = \neg Q_1 \cdot ... \cdot Q_n$
      (ii) $(Q_1 \cdot ... \cdot Q_n) \neg = Q_1 \cdot ... \cdot Q_n \neg$

tion with respect to the situation the main clause describes. This may also hold for examples such as (57a) and (57b), with *only* occurring in the exception phrase.

(57) a. ??Every man danced with two women except John with only Mary.

b. ??Every man danced with more than three women except John with only two.

Again, the Homogeneity Condition is not satisfied here. For example, the exception set in (57a) is {<John, Mary>}, and this set can neither be homogeneously excluded from nor included in every relation in (EVERY MAN, TWO WOMEN).

There are some cases where polyadic quantifiers seem to violate the Homogeneity Condition, but still are perfect with exception phrases. A case in point is the quantifier denoted by <*no man, two women*>, which allows for exception phrases:

(58) No man danced with two women except John with Mary and Sue.

The application of the Homogeneity Condition to the quantifier (NO MAN, TWO WOMEN) makes the wrong prediction. Given any model, (NO MAN, TWO WOMEN) contains a relation R just in case there are no two pairs <m, w> and <m, w′ > in R, where m is a man and w and w′ are distinct women. But this means that the set {<John, Mary>, <John, Sue>} (the only element in W([*John with Mary and Sue*])) may not be totally excluded from every relation in (NO MAN, TWO WOMEN). One element of this set may still be in a relation in the quantifer.

A plausible reason for the acceptability of (57) is the following: *two women* does not denote a quantifier ranging over individuals, but rather a quantifier ranging over *groups* of entities, namely groups consisting of two women, as in (59) (where G is an operation of group formation):

(59) [*two women*] = {X | $\exists$ xy(x $\in$ [*woman*] & y $\in$ [*woman*] & x $\neq$ y & G({x, y}) $\in$ X)}

Then, <*no man, two women*> will denote the negative universal quantifer ranging over pairs <x, y>, where x is a man and y is a group of two women. Furthermore, *John with Mary and Sue* will now denote the set of sets containing {<John, G({Mary, Sue})>} as a subset. But then the Homogeneity Condition will be satisfied.

The quantifier iterations considered above which disallowed exception phrases could be ruled out simply by applying the Homogeneity Condition. But as in the case of monadic quantifiers, the Homogeneity Condition is not sufficient. The relevant cases are parallel:

(60)   a. #Every man and some woman danced with every girl except John with Mary.
        b. #Every man and no woman danced with every girl except John with Mary.
        c. #Every man and Sue danced with every girl except John with Ann.

The Homogeneity Condition predicts that (60a-c) should be acceptable. However, the examples are ruled out because the quantifiers are not (negative) universal resumptive quantifiers meeting the Quality Condition. The sequences $<$*every man and some woman, every girl*$>$ and $<$*every man and no woman, every girl*$>$ do not denote resumptive quantifiers, and thus will not be able to take an exception phrase. In the case of (60c), the Homogeneity Condition and also the condition of being a resumptive (negative) universal quantifier are satisfied, since [$<$*every man and Sue, every girl*$>$] is a way of expressing the quantifier EVERY([*man*] $\cup$ {Sue} x [*girl*]). However, this quantifier does not have a qualitative property as its restriction, as required by (32).

The examples in (60) have shown that also in the polyadic cases, the Quality Condition plays a role in the acceptability of exception phrases besides the Homogeneity Condition: the associated quantifier, whether monadic or polyadic, has to be a universal or negative universal quantifier with a qualitative property as its restriction.

Let me summarize. The question posed at the beginning of this section has been answered in the positive as follows: precisely those quantifier iterations take exception phrases that are resumptions of universal or negative universal monadic quantifiers meeting the Quality Condition; and so these quantifiers must be of one of the forms in (55a) or (55b). But this means that, given a sequence of NPs, it is possible to predict on the basis of the monadic quantifiers denoted by the individual NPs whether this sequence denotes a polyadic quantifier that accepts exception phrases or not.

Note that the individual quantifiers in any such iteration may be expressed in different ways, for instance by a conjuncton of two universal or of two negative universal quantifiers, as in (61a) and (61b):

(61)   a. Every man and every woman danced with every boy and every girl except John with his son Bill.
        b. No man and no woman danced with any girl except John with Mary.

5.5 Polyadic exception quantifiers and reducibility

Exception phrases apply to polyadic quantifiers that are iterations of monadic quantifiers (since they are expressed by sequences of NPs denoting monadic quantifiers). Polyadic quantifiers of this sort are called 'reducible quantifiers' (cf. Keenan 1987, 1992):

(62) *Definition*

A quantifier Q of type $<n>$ is *reducible* iff, for any universe M, there are quantifiers $Q_1 \ldots Q_n$ of type $<1>$ such that $Q^M = \{R \mid \{x_1 \mid \{x_2 \mid \ldots \{x_n \mid <x_1, x_2, \ldots, x_n > \in R\} \in Q_n^M\} \ldots \in Q_2^M\} \in Q_1^M\}$

However, even though the quantifier an exception phrase applies to is generally reducible, the exception quantifier that it creates is unreducible; that is, there will always be some universe on which it cannot be defined as an iterated application of monadic quantifiers. This means that the exception quantifier could not possibly be expressed by a sequence of NPs only.

Consider the exception quantifier $Q_{MWjm} = [except]$ ($<$John, Mary$>$) ($<every\ man,\ every\ woman>$), as it is involved in (41a), repeated here as (63):

(63) Every man danced with every woman except John with Mary.

One can show that on a universe in which there are more men than John and more women than Mary, Q_{MWjm} cannot be defined as an iteration of monadic quantifiers. I will first consider exception quantifiers expressed by simple exception constructions (in which the exception phrase specifies a set of n-tuples as the exceptions):

(64) *Proposition*

For quantifiers $Q_1, \ldots, Q_n$ of type $<1>$ and an n-place relation R, the quantifier $Q = [except](R)(Q_1 \cdot \ldots \cdot Q_n)$ is not reducible, for $n > 1$.

Proof: Let A and B be sets such that $|A| > 1$, $|B| > 1$ and $a \in A$ and $b \in B$. Define $Q_{ABab}(R) = 1$ iff $A \times B \setminus \{<a, b>\} \subseteq R$ and $<a, b> \notin R$. The unreducibility of Q_{ABab} can be shown by using

Reducibility Equivalence (Keenan 1992)

For Q, Q' reducible dyadic quantifiers (of type $<2>$), $Q = Q'$ iff for any sets X and Y, $Q(X \times Y) = 1$ iff $Q'(X \times Y) = 1$.

Define $Q'(R) = 0$ for all R. Show that Q_{ABab} and Q' coincide on cross products. Let X and Y be arbitrary sets. Then it must be that $Q_{AB}(X, Y) = 0$. For assume $A \times B \setminus \{<a, b>\} \subseteq X \times Y$. Then $a \in X$. For otherwise, given the assumptions, there is an $a' \in A$, $a' \neq a$, such that $<a', b> \in A \times B$, but not $\in X \times Y$. Moreover, $b \in Y$. For otherwise there is a $b' \in B$, $b' \neq b$, such that $<a, b' > \in A \times B$, but not $\in X \times Y$. Thus $<a, b> \in X \times Y$.

So Q_{ABab} (X x Y) = 0. Thus, Q_{ABab} and Q' coincide on cross products. Now apply Reducibility Equivalence. It is not the case that Q_{ABab} = Q'. E.g., Q_{ABab}(A x B \ {<a, b>}) = 1 but Q'(A x B \ {<a, b>}) = 0. But this means that not both Q_{ABab} and Q' can be reducible. Q' clearly is reducible. E.g., define Q = {R | N({x | N({y | <x, y> ∈ R}) = 1}) = 1}, where N(P) = 0 for all sets P. Hence, Q_{ABab} must be unreducible.[14]

The proof in (64) captures only positive exception quantifiers. However, the result can straightforwardly be generalized to negative quanti-

---

[14] In order to show that a quantifier is unreducible it suffices to find one model in which the quantifier is not equivalent to an iteration of monadic quantifiers. In the model I have chosen, there are other men besides John and other women besides Mary. However, also in a universe in which John is the only man and Mary the only woman, the quantifier Q_{MWjm} is unreducible. (For this universe, (62) may sound odd, but this is besides the point.) The proof is as follows. Define Q'(R) = 1 iff a ∉ Dom(R) or b ∉ Ran(R). Let X and Y be arbitrary sets. *First case*: a ∉ X or b ∉ Y. Then <a, b> ∈ X x Y. Hence Q_{ABab}(X x Y) = 1. Furthermore, Q'(X x Y) = 1. *Second case*: a ∈ X and b ∈ Y. Then <a, b> ∈ X x Y. Hence Q'(X x Y) = 0. Furthermore, Q_{ABab}(X x Y) = 0. Thus, Q' and Q_{ABab} coincide on cross products. Q' is reducible. E.g., define Q'(R) = 1 iff Y_a({x | N_b({y | <x, y> ∈ R}) = 1}) = 1, where Y_a(A) = 1 iff a ∈ A, and N_a(A) = 1 iff a ∉ A, for any set A. Hence Q_{ABab} must be unreducible. For this universe, Q_{MWjm} in fact yields the same set of relations as the quantifier ¬(JOHN · MARY), that is, the set of all relations R which do not include the pair <John, Mary>.

However, not for all universes does the quantifier Q_{MWjm} fail to be definable as an iteration of monadic quantifiers. In any universe M_1 in which John is the only man, but there are other women besides Mary, or in any universe M_2 in which there are other men besides John, but Mary is the only woman, Q_{MWjm} can be defined as an iteration of monadic quantifiers. This can intutively be seen from the fact that (63) in M_1 would be equivalent to (1a) below, and in M_2 to (1b) below (setting aside irrelevant differences in acceptability between (63) on the one hand and (1a) and (1b) on the other hand for those universes):

(1)　a. Every man except John danced with Mary.

　　　b. John danced with every woman except Mary.

In (1a), we have an iteration of the two monadic quantifiers [*every man except John*] and [*Mary*] and in (1b) of the two monadic quantifiers [*John*] and [*every woman except Mary*]. Thus, given the relevant universes, the sequence *<every man, every woman><except John with Mary>* expresses the same quantifier as the sequences *<every man except John, Mary>* and *<John, every woman except Mary>*. The latter ones, by construction, denote iterations of monadic quantifiers.

The formal proof of the identity of the exception quantifier and the iteration of the two monadic quantifiers on M_1 (and similarly on M_2) is as follows. Show that Q_{ABab} = Y_a· Q_{Bb}, where Q_{Bb} is defined as follows: Q_{Bb}(P) = 1 iff b ∉ P and B\ {b} ⊆ P for any set P. Let X and Y be arbitrary sets. ⊆: Let Q_{ABab}(X x Y) = 1. That is, <a, b> ∉ X x Y and {a} x B \ {<a, b>} ∈ X x Y. Given the assumption, there is a c ∈ B, c ≠ b, and <a, c> ∈ X x Y. Hence a ∈ X. Then b ∉ Y. But this means that {x | Q_{Bb}({y | <x, y> X x Y}) = 1} = X, and hence Y_a· Q_{Bb}(X x Y) = 1. ⊇ : Let Y_a· Q_{Bb}(X x Y) = 1. This means that a ∈ {x | Q_{Bb}({y | <x, y> ∈ X x Y}) = 1}, which implies that B \ {b} ⊆ Y, a ∈ X, and b ∉ Y. But then <a, b> ∉ X x Y. Since A \ {a} = ∅, we have Q_{ABab}(X x Y) = 1. Q. E. D.

fiers, since (un)reducibility is preserved under internal negations (post-complements) (cf. Keenan 1992):

(65) *Proposition (Keenan 1992)*
For a generalized quantifier Q of type $<n>$, Q is (un)reducible iff $(Q\neg)$ is (un)reducible.

Moreover, even though (64) accounts only for exception quantifiers expressed by simple exception constructions, it can easily be generalized to exception constructions in which the *except*-complement denotes a generalized quantifer. As the denotations of such exception constructions have been defined in Section 4, they are unions of exception quantifiers like Q_{ABab} in (64). Now, it is a general fact that the (finite) join of such polyadic exception quantifiers again is an unreducible quantifier:

(66) *Proposition*
For quantifiers $Q_{ABab}, \ldots, Q_{AnBnanbn}$ as defined in (64) for possibly distinct $a_1, \ldots, a_n$ and $b_1, \ldots, b_n$, $Q_{A1B1a1b1}$ v $\ldots$ v $Q_{AnBnanbn}$ is unreducible.

Proof: Induction base: Let n = 2. Show $Q_{A1B1a1b1}$ v $Q_{A2B2a2b2}$ is unreducible. Use Reducibility Equivalence and a quantifier Q as defined in (64), which coincides with Q_{ABab} on cross products. Let a quantifier Q' be defined so that it coincides with $Q_{ABa'b'}$ on cross products. Show that Q v Q' coincides with Q_{ABab} v $Q_{A'B'a'b'}$ on cross products. Let P and P' be arbitrary sets. Then Q v Q'(P x P') = 1 iff Q(P x P') = 1 or Q'(P x P') = 1. But Q(P x P') = 1 iff Q_{ABab}(P x P') = 1, and Q'(P x P') = 1 iff $Q_{ABa'b'}$(P x P') = 1. But then Q v Q'(P x P') = 1 iff Q_{ABab} v $Q_{ABa'b'}$(P x P') = 1.

Induction step. Suppose $Q_{A1B1a1b1}$ v $\ldots$ v $Q_{An-1Bn-1an-1bn-1}$ is unreducible. Let Q be a quantifier which coincides with $Q_{A1B1a1b1}$ v $\ldots$ v $Q_{An-1Bn-1an-1bn-1}$ on cross products. Let Q' be defined so that it coincides with Q_{ABanbn} on cross products. Show that Q v Q' coincides with $(Q_{A1B1a1b1}$ v $\ldots$ v $Q_{An-1Bn-1an-1bn-1})$ v $Q_{A'B'a'b'}$ on cross products. Let $P_1, \ldots$, P_{n-1}, and P' be arbitrary sets. Then Q v Q'(P_1 x $\ldots$ x P_{n-1} x P') = 1 iff Q(P_1 x $\ldots$ x P_{n-1} x P') = 1 or Q'(P_1 x $\ldots$ x P_{n-1} x P') = 1. But Q(P_1 x $\ldots$ x P_{n-1} x P') = 1 iff $(Q_{A1B1a1b1}$ v $\ldots$ v $Q_{An-1Bn-1an-1bn-1})$(P_1 x $\ldots$ x P_{n-1} x P') = 1, and Q'(P_1 x $\ldots$ x P_{n-1} x P') = 1 iff $Q_{AnBnanbn}$(P_1 x $\ldots$ x P_{n-1} x P') = 1. But then Q v Q'(P_1 x $\ldots$ x P_{n-1} x P') = 1 iff $(Q_{A1B1a1b1}$ v $\ldots$ v $Q_{An-1Bn-1an-1bn-1})$ v $Q_{ABa'b'}$(P_1 x $\ldots$ x P_{n-1} x P') = 1.

6 Summary

In this paper, I have given an analysis of exception constructions which was first presented for the simplest case in which the associated quantifier of the exception phrase was a monadic quantifier and in which the exception phrase specified a particular set of entities as the exceptions. The analysis was then generalized in two steps: first, in order to account for *except*-complements denoting generalized quantifiers; and second, in order to account for polyadic quantifiers as the quantifiers an exception phrase may apply to.

Exception phrases with polyadic quantifiers present the most interesting case. They present a very clear instance of polyadic quantification in natural language. As a result on natural language expressibility, it has been shown that the exception quantifiers that polyadic exception constructions denote are not definable as iterations of monadic quantifiers and hence constitute unreducible polyadic quantifiers.

References

Barwise, Jon, and Robin Cooper. 1981. Generalized Quantifiers and Natural Language. *Linguistics and Philosophy* 4.

van Benthem, Johan. 1989. Polyadic Quantification. *Linguistics and Philosophy* 10.

Clark, Robin, and Ed Keenan. 1985/6. The Absorption Operator. *The Linguistic Review* 5.

van Eijck, Jan. 1991. Partial Quantifiers. In *Generalized Quantifier Theory and Applications*, ed. Jaap van der Does and Jan van Eijck. Amsterdam: Dutch Network for Language, Logic and Information.

Fine, Kit. 1977. Properties, Propositions, and Sets. *Journal of Philosophical Logic* 6.

von Fintel, Kai. 1993. Exceptive Constructions. *Natural Language Semantics* 1.

Higginbotham, Jim, and Robert May. 1981. Questions, Quantifiers and Crossing. *The Linguistic Review* 1.

Hoeksema, Jack. 1987. The Logic of Exception. In A. Miller (ed.). *Proceedings of ESCOL* 4. Columbus: The Ohio State University.

Hoeksema, Jack. 1989. Exploring Exception Phrases. In *Proceedings of the Seventh Amsterdam Colloquium*, ed. L. Torenvliet and M. Stokhof. Amsterdam: ITLI.

Hoeksema, Jack. 1991. The Semantics of Exception Phrases. In *Generalized Quantifier Theory and Applications*, ed. Jaap van der Does and Jan van Eijck. Amsterdam: Dutch Network for Language, Logic and Information.

Keenan, Ed. 1987. Unreducible N-ary Quantifiers. In *Generalized Quantifiers: Linguistic and Logical Approaches*, ed. Peter Gaerdenfors. Dordrecht: Reidel.

Keenan, Ed. 1992. Beyond the Frege Boundary. *Linguistics and Philosophy* 15.

May, Robert. 1989. Interpreting Logical Form. *Linguistics and Philosophy* 12.

Moltmann, Friederike. 1992. *Coordination and Comparatives*. Ph D thesis MIT, Cambridge (Mass.).

Moltmann, Friederike. To appear. Exception Sentences and Polyadic Quantification. *Linguistics and Philosophy*.

Muskens, Reinhart. 1989. *Meaning and Partiality*. Ph D thesis, University of Amsterdam, Amsterdam.

Nam, Seungho. 1991. N-ary Quantifiers and the Expressive Power of NP-Composition. In *Generalized Quantifier Theory and Applications*, ed. Jaap van der Does and Jan van Eijck. Amsterdam: Dutch Network for Language, Logic and Information.

Srivastav, Veneeta. 1991. *Wh-Dependencies in Hindi and the Theory of Grammar*. Ph D thesis, Cornell University, Ithaca (N.Y.).

Westerståhl, Dag. 1992. *Iterated Quantifiers*. ITLI Prepublication Series, University of Amsterdam, Amsterdam.

7

(In)definites and genericity

HENRIËTTE DE SWART

In this paper I develop a unified analysis of generic and non-generic readings of definite and indefinite NPs in sentences expressing characteristic predication. I build on examples from English and French in order to show that genericity is not cross-linguistically related to the presence of indefinite NPs but can be expressed by means of definite NPs as well. An analysis of adverbs of quantification as generalized quantifiers over events combined with an interpretation of indefinite NPs as dynamic existential quantifiers and of definite NPs as context-dependent quantifiers yields the right interpretation of generic sentences.

1 Varieties of genericity

1.1 Introduction

The expression of generalizations in natural language is not a unified phenomenon. Krifka et al. (1995) distinguishes reference to *kinds* (1) and *characteristic predication* which generalizes over individuals (2):

(1) a. The potato was first cultivated in South America.
    b. Potatoes were introduced here by the end of the 17th century.

(2) a. A potato contains vitamin C, amino acid, protein and thiamin.
    b. Italians drink wine with their dinner.

Reference to kinds and characteristic predication have certain properties in common, but there are also a number of linguistic differences between them. Typically, in English at least, we use definite NPs or bare plurals to refer to kinds, whereas indefinite singulars or bare plu-

I wish to thank Leonie Bosveld-de Smet, Cleo Condoravdi, Emiel Krahmer, my co-editors and the anonymous reviewer for helpful comments on an earlier version. The research for this paper was supported by a fellowship of the Royal Netherlands Academy of Arts and Sciences (KNAW), which is hereby gratefully acknowledged.

Quantifiers, Deduction, and Context
Makoto Kanazawa, Christopher Piñón, and Henriëtte de Swart, editors

rals are most frequent in characteristic sentences. Singular indefinites are excluded in contexts in which kind-level predication is involved:

(3) a. The dodo is extinct.
   b. *A dodo is extinct.

In contradistinction to (3a), (3b) is unacceptable, unless we assign the sentence a taxonomic reading and claim that a particular subspecies of dodos is extinct. Observations like these suggest that the two phenomena ought to be treated in different ways (see Krifka et al. 1995). My main concern will be with generic and non-generic interpretations of indefinite and definite NPs in the expression of characteristic predication in English and French.

The kind of genericity which plays a role in characterizing sentences is a property of sentences rather than NPs. Indefinite singulars cannot be kind-referring (3b), but they can certainly get a generic reading in a characterizing sentence (2a). Therefore, the source of genericity is not in the indefinite NP itself. This has led many authors to introduce a generic operator, which is not phonologically realized, but is closely related to adverbs of quantification like *always*, *usually*:

(4) a. Italians smoke.
   b. Italians usually smoke.

Both the adverbs and the generic operator can be treated as two-place quantifiers, relating a restrictor and a matrix:

Q [Restrictor] [Matrix]

For reasons of exposition, I will use explicit adverbs, rather than rely on a phonologically null generic operator. With Rooth (1985), I assume that these adverbs associate with focus, and quantify over the part of the sentence that provides the background to the focus. Material which is part of the background thus ends up in the restrictor, material which is in focus in the matrix.

1.2   Existential and generic readings of indefinites

The bare plural in (4) is clearly a generic NP: we express a generalization about Italians. This does not mean that any indefinite which occurs in a generic sentence is a generic NP. A well-known counterexample is the kind of sentence in (5) (originally from Milsark 1974), which has the readings listed under (5a) and (5b):

(5) Typhoons often arise in this part of the Pacific.
   a. Typhoons have a common origin in this part of the Pacific.
   b. There arise typhoons in this part of the Pacific.

The (a)-reading expresses a generalization about typhoons, and the bare plural is intuitively characterized as a generic NP. In the (b)-

reading, the bare plural has an existential reading, and the generalization expressed holds for this part of the Pacific:

(6) a. OFTEN [typhoons] [arise in this part of the Pacific]
   b. OFTEN [in this part of the Pacific] [typhoons arise]

Indefinite singulars and bare plurals get an existential reading if they are in focus and therefore part of the matrix, and a generic reading if they are background and part of the restrictor.

We find more cases of non-generic indefinite NPs in generic sentences if we extend our discussion to NPs in object position. (7) is ambiguous and has (at least) two readings:

(7) Anne always knits Norwegian sweaters
   a. ALWAYS [Anne knits some kind of sweaters]
      [Anne knits Norwegian sweaters]
   b. ALWAYS [Anne knits something] [Anne knits Norwegian sweaters]

(7a) allows the inference that all sweaters that Anne knits will be Norwegian sweaters. In (7b), all the things she knits will be Norwegian sweaters. We expect the bare plural in the (b)-reading to be existential, because it ends up in the matrix. Although the indefinite NP *sweaters* is part of the restrictor in the (7a)-reading, the bare plural also gets an existential, rather than a generic interpretation. (7b) is not a generalization over sweaters claiming that sweaters are Norwegian sweaters when knitted by Anne, but a generalization over situations in which Anne knits some kind of sweater. This suggests that only a subset of the indefinite NPs that are part of the restrictor gets a generic interpretation. Examples which confirm this hypothesis are the ones in (8a) and (9a), which get the representations in (8b) and (9b) respectively:

(8)   a. A woman who has a cat usually likes it.
      b. USUALLY [a woman who has a cat$_i$] [likes it$_i$]
(9)   a. If a drummer lives in an apartment building, it is usually half empty.
      b. USUALLY [a drummer lives in an apartment building$_i$]
         [it$_i$ is half empty]

(8a) expresses a generalization over cat-owning women, not over cats owned by women. That is, the two indefinites are treated asymmetrically: the cats are dependent on the women, and *a cat* gets an existential interpretation; only the NP *a woman* is generic in this case. (9a) is similar: we generalize over apartment buildings lived in by drummers, rather than over drummers who live in an apartment building. So *a drummer* is existentially quantified, whereas the NP *an apartment building* gets a generic interpretation.

These examples are instantiations of 'asymmetric' quantification, which involve an asymmetry between a 'head' variable and a 'dependent' variable. They show that the focus/background distinction is insufficient to account for generic and non-generic interpretations of indefinite NPs. We need to subdivide the restrictor into a part which tells us what the quantification is 'about', and which I will call the *topic*, and a part which provides additional information on the topic.[1]

1.3 The French paradigm

In order to sharpen our intuitions about generic and non-generic readings of indefinites in generic sentences, I will discuss similar sentences in French. Consider (10), which is the counterpart to (2):

(10)   a. Un Italien boit généralement du vin à table.
           An Italian drinks generally INDEF-MASS wine at table
      b. *Des Italiens boivent généralement du vin à table.
           INDEF-PL Italians drink generally INDEF-MASS wine at table
      c. Les Italiens boivent généralement du vin à table.
           DEF-PL Italians drink generally INDEF-MASS wine at table

In a Romance language such as French, characteristic predication is expressed by indefinite singulars or definite plurals. Indefinite plurals cannot be used to provide the NP we generalize over as (10b) shows.[2]

Other than in generic contexts, the indefinite plural *des N* behaves pretty much like the English bare plural. It allows discourse anaphora and donkey anaphora, as in (11):

(11)   a. Hier soir, *des terroristes basques$_i$* ont essayé d'enlever le Premier Ministre. *Ils$_i$* n'ont pas eu de chance: *ils$_i$* ont été arrêtés ce matin.
           Yesterday evening, *Basque terrorists* tried to kidnap the Prime Minister. *They* were not very lucky: *they* have been arrested this morning.

---

[1] This subdivision is reminiscent of Vallduví's (1992) analysis of information packaging, and his division of the (back)ground into a *link* (which I call the topic) and a *tail* (the additional background information on the topic). I will not use Vallduví's terminology here, because he holds a slightly different view on how information packaging relates to association with focus.

[2] In colloquial French, there is an alternative construction in which *des N* occurs in relation with so-called generic *ça*:

(i) Des chiens/ Les chiens/ Un chien, ça aboie.
    INDEF-PL dogs/ DEF-PL dogs/ INDEF-SG dog, that bark
(ii) Beaucoup d'arbres, ça attire les insectes.
    Many trees, that attracts insects

This construction is quite different from the ones treated in this paper and I will present no analysis of sentences like these. See Auger 1993 for discussion.

b. Tous les paysans qui ont *des ânes têtus*$_i$ les$_i$ battent.
All farmers who have *stubborn donkeys* beat them.

The rule of thumb is: bare plurals that have an existential reading translate as *des* N in French, generic bare plurals translate as *les* N. This suggests that we can use the contrast between *des* N and *les* N as a diagnostic test to determine which NPs in a sentence are generic. In the French translation of a sentence like (5), the ambiguity is resolved:

(12)   a. Les ouragans violents naissent souvent dans cette partie du Pacifique.
DEF-PL typhoons often arise in this part of the Pacific

b. Dans cette partie du Pacifique naissent souvent des ouragans violents.
In this part of the Pacific arise often INDEF-PL typhoons

(12a) corresponds with the interpretation in which typhoons in general arise in this part of the Pacific, whereas (12b) only has the interpretation in which there arise typhoons in this part of the Pacific. *Des* N can occur in a generic sentence, if it is focal and part of the matrix, because there it will get an existential reading. The NP about which a certain generalization is stated cannot be *des* N, so if the topic is typhoons, we switch to a definite plural, as in (12a). The French counterpart to (7) shows that *des* N can even be part of the restrictor on the quantifier, as long as it need not be generic:

(13) En général, Anne tricote des chandails norvégiens.
In general, Anne knits INDEF-PL Norwegian sweaters

(13) is ambiguous in the same way as (7) is, and it has the readings represented in (7a) and (7b). Given that *des* N is never generic, this confirms our intuition that in neither of the two interpretations is the indefinite plural generic in nature: the generalization stated is not in any way 'about' sweaters and *des* N is not the topic of the sentence. Nevertheless, it is clear from the representation in (7a) that the indefinite NP is part of the restrictor and not just of the matrix.

2   Bound variable readings

The French data confirm my claim about the interpretation of indefinite NPs in generic sentences. Focal indefinites occur in the matrix, where they get an existential reading. Generic indefinites are a (not necessarily proper) subset of the indefinite NPs that are interpreted as part of the restrictor. The only indefinites which can be properly called generic NPs are those which count as topics: they tell us for which objects the generalization hold. The analysis I will develop is based on the intuition that the NPs which we described as properly generic are somehow 'bound' by the quantifier. Generic readings are then described as 'bound variable' readings.

The predominant view in the field right now (Krifka et al. 1995) seems to be that bound variable readings ought to be handled by analyzing indefinites as variables and adopting one version or another of the unselective binding theory. However, given that *des* N does not behave as a bound variable in relevant contexts like (10), (12), and (13), it would be preferable to translate such an indefinite NP by means of an existential quantifier. Therefore, I will adopt a more indirect approach to bound variable readings by interpreting indefinite NPs uniformly as (dynamic) existential quantifiers. In this analysis, bound variable readings come out as an instance of 'pseudo-binding', a notion which will be made precise below. This approach will be shown to have the advantage of providing a natural extension to generic readings of definite NPs.

2.1 Dynamic existential quantifiers

Groenendijk and Stokhof (1991, 1992) introduce dynamic existential quantifiers which differ from their static counterparts of classical predicate logic in that they provide an "anchor" for variable assignments in subsequent sentences. This allows for binding relations beyond the sentence boundary. Interpreting the indefinite NP as a dynamic existential quantifier gives (14) the representation under (14a). It is translated as in (14b), which is rewritten as (14c):

(14) A dog came in. It lay down under the table.

   a. $[\mathcal{E}d \, [\uparrow \text{Dog}(d)] \; ; \; \uparrow \text{Come-in}(d)] \; ; \; \uparrow \text{Lay-down}(d)]$

   b. $[\lambda p \, \exists x \, [\text{Dog}(x) \wedge \text{Come-in}(x) \; \wedge \; \{x/d\}^{\vee}p]^{\wedge}(\uparrow \text{Lay-down}(d))]$

   c. $[\lambda q \, \exists x \, [\text{Dog}(x) \wedge \text{Come-in}(x) \wedge \text{Lay-down}(x) \; \wedge \; \{x/d\}^{\vee}q]]$

The up-arrow ($\uparrow$) signals the dynamic character of the proposition: the dynamic meaning or context change potential of a sentence is taken to be its ability to constrain subsequent discourse. (14b) shows that in the first sentence the variable assignment anchors the discourse marker d to the individual x. This variable assignment is carried on to the next sentence, which is related to it by means of dynamic conjunction: $\{x/d\}$ requires subsequent occurrences of d to also be attached to x. The pronoun *it* can then be interpreted as referring to the same entity x. This accounts for both discourse and donkey anaphora.

Chierchia (1992) introduces the notion of dynamic generalized quantifier, which establishes a relation between dynamic properties. The definition of dynamic and conservative generalized quantifiers is in (15):

(15) $\mathbf{D}^{+}(P)(Q) \; = \uparrow D \, (\lambda x \downarrow {}^{\vee}P(x)) \, (\lambda x \downarrow [{}^{\vee}P(x) \; ; \; {}^{\vee}Q(x)])$

P and Q are dynamic properties, which have an additional place holder for discourse continuations. The downarrow ($\downarrow$) brings us back from context change potentials to type t expressions. $\mathbf{D}^+$ is defined in terms of its static counterpart D, which relates sets of individuals as in standard generalized quantifier theory. The context change potential of a donkey sentence is represented as follows:

(16) Every farmer who owns a donkey beats it.

    a. **Every**$^+$ (x is a farmer that owns a donkey)(x beats it) =
        $\uparrow$ EVERY ($\lambda x \downarrow [x$ is a farmer that owns a donkey])
        ($\lambda x \downarrow [x$ is a farmer that owns a donkey *and* beats it])

    b. $\uparrow$ EVERY $[\lambda x \downarrow [\uparrow$ Farmer(x) ; $\mathcal{E}d [\uparrow$ Donkey(d) ; $\uparrow$ Own$(x, d)]]]$
       $[\lambda x \downarrow [\uparrow$ Farmer(x) ; $\mathcal{E}d [\uparrow$ Donkey(d) ; $\uparrow$ Own$(x, d)]$;
       $\uparrow$ Beat$(x, d)]]$

    c. $\uparrow \forall x [[$Farmer$(x) \wedge \exists y [$Donkey$(y) \wedge$ Own$(x, y)]] \rightarrow$
      $\exists y [$Donkey$(y) \wedge$ Own$(x, y) \wedge$ Beat$(x, y)]]$

The conservative dynamic quantifier **Every**$^+$ relates the dynamic property of 'farmers that own a donkey' to 'farmers that own a donkey *and* beat it'. The binding capacities of the dynamic existential quantifier extend to the dynamically conjoined 'beat it'.

    I will assume that Q-adverbs are interpreted similarly as dynamic conservative generalized quantifiers over (minimal) events or situations. The general definition is given in (17), which is like (15), except that the adverbial quantifier A now ranges over events:

(17) $\mathbf{A}^+(P)(Q)$ $= \uparrow A (\lambda e \downarrow {}^{\vee}P(e)) (\lambda e \downarrow [{}^{\vee}P(e)$; ${}^{\vee}Q(e)])$

P and Q are dynamic properties, which have an additional place holder for discourse continuations. The downarrow ($\downarrow$) brings us back from context change potentials to type t expressions. The conservative adverbial quantifier $\mathbf{A}^+$ is defined in terms of its static counterpart A, which relates sets of events as in standard generalized quantifier theory. This leads to the following representation of example (18):

(18) When John is in the bathtub, he always sings.

    a. **Always**$^+$ (e is an event of John in the bathtub)
       (e is an event of John singing) =
       $\uparrow$ ALWAYS $[\lambda e \uparrow$ In-bath$(j, e)] [\lambda e \uparrow$ In-bath(j, e) ; $\uparrow$ Sing$(j, e)]$

    b. $\uparrow \forall e [[$In-bath$(j, e)] \rightarrow [$In-bath$(j, e) \wedge$ Sing$(j, e)]]$

In quantified sentences introduced by an *if*- or a *when*-clause, the restriction on the adverbial quantifier is provided by the subordinate clause, so (18) gets the interpretation that all events of John being in the bathtub are such that they are events of John singing in the bathtub.[3] The treatment extends to sentences containing an indefinite in a straightforward way, namely by iteration of bindings:

(19) When John invites a friend, he always cooks dinner for her.

    a. **Always**$^+$ (e is an event of John inviting a friend)
       (e is an event of John cooking dinner for her) $=$
       $\uparrow$ ALWAYS $[\lambda e\, \mathcal{E}d\, [\uparrow \text{Friend}(d, j)\, ;\, \uparrow \text{Invite}(j, d, e)]]$
       $[\lambda e\, \mathcal{E}d\, [\uparrow \text{Friend}(d, j)\, ;\, \uparrow \text{Invite}(j, d, e)]\, ;\, \uparrow \text{Cook-for}(j, d, e)]]$

    b. $\uparrow \forall e\, [[\exists x\, \text{Friend}(x, j) \land \text{Invite}(j, x, e)] \rightarrow$
      $[\exists x\, \text{Friend}(x, j) \land \text{Invite}(j, x, e) \land \text{Cook-for}(x, j, e)]]$

Each alternative event generated by the *when*-clause encodes a choice of a friend as the value of the indefinite *a friend*. Because of the quantification over events, there is no need to suppress the existential import of indefinites within the restrictive clause. The general mechanism of carrying forward the value assigned to the indefinite guarantees the desired anaphoric binding relation.

2.2 Quantification over events and individuals

An objection which has been raised against the event-based analysis of Q-adverbs concerns non-episodic sentences such as (20):

(20) When a cat has blue eyes, it is often intelligent.

According to Lewis (1975), we may wonder whether it even makes sense to talk about quantification over events here, because a property such as *having blue eyes* is an individual-level property which is not bound to particular occasions. The unacceptability of (21), which contrasts with (18) provides a strong argument in favor of the view that what (20) really quantifies over is cats:

(21) *When Minouche has blue eyes, it is often intelligent.

Given that the color of a cat's eyes is independent of a particular occasion, we need an indefinite NP to make the Q-adverb operate non-vacuously. This has led to the suggestion (Kratzer 1995) that only stage-level predicates as in (18) and (19) come with a Davidsonian event argument, whereas individual-level predicates such as the one in (20)/(21) do not have a spatio-temporal location argument. In De Hoop and De Swart (1990) and De Swart (1991) it has been argued that this line of reasoning cannot be correct, for it would make it hard to account for similar contrasts in the following examples:

(22) a. *When Anil died, his wife usually killed herself.
    b. When an Indian died, his wife usually killed herself.

---

[3]Strictly speaking, the events of John being in the bathtub and of him singing are not to be treated as identical events, but as two distinct, but overlapping events. See De Swart 1991 and 1995 for a more sophisticated treatment of such cases.

(23) a. *When Mary built Jim's house, she always built it well.

b. When Mary built a house, she always built it well.

It is quite clear why the (a)-sentences are unacceptable: although the predicate is stage-level, the same action cannot be repeated with respect to the same individual. That is, they are 'once-only' predicates. An indefinite NP in the subject/object position is needed in order to create a plurality of situations the adverb can quantify over. A unified analysis of the cases in (20)-(23) can be developed if we assume that individual-level predicates are a special case of 'once-only' predicates. They introduce a Davidsonian event variable just like regular stage-level predicates, but they come with the following presupposition:

(24) *Uniqueness presupposition on the Davidsonian argument*
If not empty, the set of events that is in the denotation of a 'once-only' predicate is a singleton set for all models and each assignment of individuals to the arguments of the predicate.

With respect to a particular assignment function, the proposition denotes a singleton set of events. The generalization for both once-only and individual-level predicates is that Q-adverbs do not quantify over singleton sets. They need a plurality of situations, otherwise the quantification is in some sense trivial.[4] This rules out the unacceptable sentences in (21), (22a), and (23a): their predicates satisfy the uniqueness presupposition on the Davidsonian argument and they do not allow quantification, because there is no plurality of situations. (20), (22b), and (23b) are all right, because the indefinite NP creates a plurality of situations. Compare the representation of (22b) in (25):

(25) When an Indian died, his wife usually killed herself.

$\uparrow$USUALLY $[\lambda e\, \exists x\, \text{Indian}(x) \wedge \text{Die}(x, e)]$

$[\lambda e\, \exists x\, \text{Indian}(x) \wedge \text{Die}(x, e) \wedge \text{the } y\, \text{Wife-of}(y, x) \wedge \text{Kill}(y, y, e)]$

The main difference between (19) and (25) resides in the relation between individuals and events. In (19), John can invite the same friend several times, so indirect binding of the existential quantifier does not induce a bound variable reading on the indefinite NP. In (25), quantification over events is indistinguishable from quantification over individuals. The reason is that individuals die only once, so there is a

---

[4]The formulation in (24) involves universal quantification over assignment functions and models. Thus we do not expect quantification to be anomalous in (i):

(i) There is only one man in the room and he knows French. Thus, every man in the room knows French

The fact that we have a singleton set in (i) is accidental, and does not have consequences for the interpretation of the sentence. (i) contrasts with (ii), which is anomalous because we do not understand people to have more than one mother:

(ii) *Every mother of Susan brought a present

uniqueness presupposition on the Davidsonian argument. As a result, the set of (minimal) events in which an Indian dies is in one-one correspondence with the set of dying Indians. There are as many dying Indians as there are (minimal) events in which an Indian dies:

(26) $|\{e|\exists x \text{ Indian}(x) \wedge \text{Die}(x,e)\}| = |\{x| \text{Indian}(x) \wedge \exists e \text{ Die}(x,e)\}|$

This induces a bound variable reading on the indefinite NP: it is quasi-bound along with the event variable. In the event-based analysis of adverbs of quantification we have adopted, there is of course no real binding of the individual variable, because quantification is over events. The meaning effect has nothing to do with unselective binding; it is just a consequence of the pragmatics of the sentence.

We can handle examples involving individual-level predicates in the same way, for they share the uniqueness presupposition with once-only predicates. For instance:

(27) When a cat has blue eyes, it is often intelligent.
$\uparrow$ OFTEN $[\lambda e\, \exists x \text{ Cat}(x) \wedge \text{Have-blue-eyes}(x,e)]$
$[\lambda e\, \exists x \text{ Cat}(x) \wedge \text{Have-blue-eyes}(x,e) \wedge \text{Intelligent}(x,e)]$

Given that individual-level predicates are once-only, no two (minimal) events in which there is a cat with blue eyes will involve the same cat. Every event has its own cat, so the set of these events is in one-one correspondence with the set of cats having blue eyes. This explains why the topic of the sentence is blue-eyed cats, although, formally, quantification is about events containing a blue-eyed cat. The process in which binding of the event variable has the effect of indirect quantification over individuals because of the once-only character of the predicate will be referred to as 'pseudo-binding'.

In English, pseudo-binding does not only apply to singular indefinites (28a), but also to bare plurals as in (28b):

(28) a. An Italian usually drinks wine at dinner.
b. Italians usually drink wine at dinner.

The property of drinking wine at table is distributively predicated of Italians in general, so (28b) ends up as equivalent to (28a).

2.3 Existential disclosure

It is interesting to compare the pragmatic account in terms of pseudo-binding to the semantic analysis of bound variable readings developed by Chierchia (1992) and Dekker (1993). In order to give a semantic account of the bound variable reading in a framework which treats indefinite NPs as (dynamic) existential quantifiers, we need a mechanism to turn indefinite NPs into variables. Type-shifting mechanisms which realize this are discussed by Partee (1987). A good candidate would be Partee's *ident* or BE operation. *ident* is defined in (29a) and turns type e expressions into type $\langle e, t \rangle$ expressions by means of identity and λ-abstraction:

(29)　　a. *ident* : $a \rightarrow \lambda x\,[x = a]$
　　　　b. BE: $\lambda \mathcal{P}\,\lambda x\,[\mathcal{P}(\lambda y\,[y = x])]$

BE is an operator which maps type $\langle \langle e, t \rangle, t \rangle$ expressions onto type $\langle e, t \rangle$ expressions, with the semantics in (29b). Partee shows that $\mathrm{BE}(a(\mathrm{P})) = \mathrm{P}$:

(30)　　a. $\|\,a\,\mathrm{P}\,\| = \lambda Q\,[\exists x\,[P(x) \wedge Q(x)]]$
　　　　b. $\|\,\mathrm{BE}\,a\,\mathrm{P}\,\| = \lambda \mathcal{P}\,\lambda x\,[\mathcal{P}\,(\lambda y\,[y = x])]\,(\lambda Q\,[\exists x\,[P(x) \wedge Q(x)]])$
　　　　c. $\|\,\mathrm{BE}\,a\,\mathrm{P}\,\| = \lambda x\,P(x)$

In (30b), we apply the type-shifting operator BE to the generalized quantifier denotation of *a* N, as defined in (30a). After λ-conversion, we end up with (30c), which is just the denotation of N. In this way, we map existentially quantified NPs onto the set of individuals which satisfy the property referred to by the common noun.

In DMG, a type-shifting mechanism much like Partee's *ident* or BE operation is defined and called 'existential disclosure' (Dekker, 1993; Chierchia, 1992). Existential disclosure transforms an existentially quantified NP into a predicate by means of identity and λ-abstraction. For instance, the antecedent of (25) repeated here as (31a) denotes a set of events. (31a) can be transformed into (31c), which denotes a set of pairs of an event and an individual via conjunction with the identity statement in (31b):

(31)　　a. $[\lambda e\,\mathcal{E}d\,[\uparrow \mathrm{Indian}(d)\,;\,\uparrow \mathrm{Die}(d, e)]]$
　　　　b. $[\lambda e\,\lambda y\,\mathcal{E}d\,[\uparrow \mathrm{Indian}(d)\,;\,\uparrow \mathrm{Die}(d, e)]\,;\,\uparrow [d = y]]$
　　　　c. $[\lambda e\,\lambda y\,[\mathrm{Indian}(y) \wedge \mathrm{Die}(y, e)]]$

The Q-adverb accordingly shifts from a quantifier over events to a quantifier which binds pairs of events and individuals. In (32), the predicate is so loosely tied to occasions that we may be tempted to drop the event argument altogether and reduce (32c) further to (32d):

(32) When a cat has blue eyes, it is often intelligent.
　　　　a. $[\lambda e\,\mathcal{E}d\,[\uparrow \mathrm{Cat}(d)\,;\,\uparrow \mathrm{Have\text{-}blue\text{-}eyes}(d, e)]]$
　　　　b. $[\lambda e\,\lambda y\,\mathcal{E}d\,[\uparrow \mathrm{Cat}(d)\,;\,\uparrow \mathrm{Have\text{-}blue\text{-}eyes}(d, e)]\,;\,\uparrow [d = y]]$
　　　　c. $[\lambda e\,\lambda y\,[\mathrm{Cat}(y) \wedge \mathrm{Have\text{-}blue\text{-}eyes}(y, e)]]$
　　　　d. $[\lambda y\,[\mathrm{Cat}(y) \wedge \mathrm{Have\text{-}blue\text{-}eyes}(y)]]$

Reduction of (32c) to (32d) is not formally legitimated, but is dependent on world knowledge. With individual-level predicates, the one-one correspondence between sets of individuals and sets of events can pragmatically license such inferences.

Chierchia suggests that existential disclosure gives us the topic that is, what the quantification is about. This seems to be a suitable way to restrict the application of this operation, because generic NPs are topics in the sense that the generalization expressed by the sentence is 'about' that NP.[5] The operation extends in a natural way to cases like (33), in which the restriction is not given by an *if*- or *when*-clause, but via association with focus (Krifka 1995):

(33) An intelligent cat usually likes milk.
   a. $[\lambda e\, \exists X\, \mathcal{E}d\, [\uparrow \text{Cat}(d)\ ;\ \uparrow \text{Intelligent}(d,e)\ ;\ \uparrow X(d,e)]]$
   b. $[\lambda e\, \lambda y\, \exists X\, \mathcal{E}d\, [\uparrow\text{Cat}(d)\ ;\ \uparrow \text{Intelligent}(d,e)\ ;\ \uparrow X(d,e)]\ ;\ \uparrow [d=y]]$
   c. $[\lambda e\, \lambda y\, \exists X\, [\text{Cat}(y) \wedge \text{Intelligent}(y,e)\ \wedge\ X(y,e)]]$
   d. $[\lambda y\, [\text{Cat}(y) \wedge \text{Intelligent}(y)\ \wedge\ X(y,e)]]$

Reconstructing Krifka's (1995) analysis of focus in generic sentences in an event-based semantics gives us (33a) as a representation of the antecedent. The variable X stands for a contextually determined alternative of liking milk. If suitable alternatives are chosen among a set of closely related properties, this means that only individual-level predicates will be able to fill this position (for instance, hating milk, liking yogurt). We have the same situation then as we had in (32), and we can reduce the antecedent to quantification over intelligent cats by dropping reference to the event variable of the restrictor altogether.

In an event-based approach to Q-adverbs, existential disclosure is undesirable because it forces us to type-shift the quantifier and turn it into an operator over pairs of events and individuals. Furthermore, Chierchia and Krifka need similar pragmatic constraints as the pseudo-binding approach. Crucially, as pointed out with respect to examples (32) and (33) above, the reduction of quantification over pairs of individuals and events to quantification over individuals simpliciter is not formally licensed, but is dependent on pragmatic considerations. We must assume that existential disclosure only applies in cases where the set of events is in one-one correspondence with the set of individuals which constitutes the topic of the generalization. If we need these pragmatic constraints anyhow, there is no reason why an event-based semantics cannot be a viable alternative to an unselective binding approach.[6] In the remainder of this paper, I will therefore maintain an interpretation of Q-adverbs as generalized quantifiers over events.

---

[5]Partee uses type-shifting to interpret predicative constructions such as:
   (i) Susan is a lawyer
Predicative NPs usually provide new information, and the indefinite *a lawyer* in (i) is not a topic. So the constraints on existential disclosure in generic sentences are different from those in predicative BE-constructions. The fact that *des* N occurs in predicative sentences is thus unrelated to the fact that indefinite plurals in French do not usually have generic readings.

[6]Another objection which has been made to quantification over situations concerns the interpretation of so-called 'bishop' sentences such as (i):
   (i) When a bishop meets a man, he always blesses him.

2.4 French *des* N

The generalization to make here is that the contexts in which pseudo-binding applies are the cases in which we get bound variable readings of the indefinite NP, and in which the NP is truly generic. This allows us to formulate an appropriate constraint on the interpretation of the French indefinite plural *des* N. Just like other indefinite NPs, it denotes a dynamic existential quantifier. However, unlike the indefinite singular *un* N, the English indefinite singular *a* N and the bare plural, *des* N does not allow pseudo-binding in the presence of an adverb of quantification. This rules out examples like (34a) and (35a):

(34) a. *En général, des Indiens meurent jeunes.
          In general, INDEF-PL Indians die young
      b. En général, les Indiens meurent jeunes.
          In general, DEF-PL Indians die young

(35) a. *Des Italiens boivent généralement du vin à table.
          INDEF-PL Italians drink usually wine at table
      b. Les Italiens boivent généralement du vin à table.
          DEF-PL Italians drink usually wine at table

The two predicates in (34) and (35) are once-only. This means that the only way to satisfy the plurality condition on quantification would be to pseudo-bind the individual variable instead, but as we observed earlier, *des* N does not allow this.

Although *des* N is excluded from contexts like (34) and (35), pseudo-binding of indefinite plurals is allowed under the influence of modal operators with a strong deontic or prescriptive flavor as in (36a):

(36) a. Des agents de police ne se comportent pas ainsi dans une
          situation d'alarme.
          INDEF-PL police officers do not behave like that in an emer-
          gency situation
      b. Les agents de police ne se comportent pas ainsi dans une
          situation d'alarme.
          DEF-PL police officers do not behave like that in an emer-
          gency situation

As Carlier points out, (36a) would be uttered to reproach a subordinate with his behavior. (36b) does not have the same normative value, but gives us a descriptive generalization which could possibly be refuted by providing a counterexample. It seems that *des* N only allows pseudo-binding with a specialized interpretation.

My explanation of this pattern is diachronic in nature. I assume that the restrictions on pseudo-binding are due to the origin of *des*

See De Swart (to appear) for a reply to this argument.

as a partitive determiner, meaning 'some but not all'. The notion of partitivity is incompatible with a bound variable reading, because pseudo-binding has the effect that the existential quantification indirectly amounts to a quasi-universal, generic interpretation, and the set of individuals denoted by the common noun becomes available as a whole. Thus, the original, partitive-like *des* cannot get a generic reading. Over the years, *des* has lost most of its partitive character, and in modern French it behaves like a regular weak, indefinite NP. One indication is that pseudo-binding is now allowed in modal contexts like (36a). It still doesn't show up in contexts like (34) and (35), though. This can be considered as a remnant of an earlier stage of the language. Possibly, future developments will allow pseudo-binding even in the context of adverbs of quantification. However, this is not necessarily what we need to expect. We know that in general, not all indefinite NPs get generic readings, even in a language like English. Numerals such as *two* N allow a collective generic reading as in (37a), but they usually get a partitive reading in the subject position of an individual-level predicate (37b). Similarly, *many/few* N are interpreted proportionally rather than generically in such contexts as (37c):

(37) a. Two black cards are worth more than one red one.
b. Two students like semantics (= two of the students)
c. Many students like semantics (= a large proportion)

Just like numerals, *many* and *few*, *des* is not really suitable for the expression of genericity, but it does take up a partitive interpretation in the presence of an individual-level predicate as the following example (borrowed from Galmiche, 1986) shows:

(38) Des fauteuils sont bancals.
INDEF-PL armchairs are shaky
Some (of the) armchairs are shaky

According to Galmiche, the plural indefinite selects a subset of the contextually relevant armchairs as the referent of the NP, so the sentence means 'some of the armchairs are shaky'. This reasoning pattern is confirmed by an observation Carlier (1989) makes, namely that *des* N has a taxonomical reading in contexts like (39):

(39) a. Un colibri peut voler en arrière
INDEF-SG humming-bird can fly backwards
b. Des colibris peuvent voler en arrière
INDEF-PL humming-birds can fly backwards
c. Beaucoup de colibris peuvent voler en arrière
Many of humming-birds can fly backwards

(39a) describes a property of humming-birds in general. *Des* excludes this generalization pattern and (39b) can only mean that certain subspecies of humming-birds are capable of flying backwards. We can give a similar interpretation to (39c), which quantifies either over individual humming-birds or over sub-species of humming-birds. The data in (38) and (39) suggest that the present-day behaviour of *des* N is in fact closer to that of indefinite NPs involving numerals and *many*, *few* than to singular indefinites and bare plurals.

We could discuss at length the question why in present-day French pseudo-binding of indefinite plurals is possible in deontic contexts such as (36a), but not in the presence of an adverb of quantification as in (34a) and (35a), and why *des* N gets a partitive rather than a generic interpretation in the subject position of individual-level predicates (38), but I think such a discussion would miss the point. Historically, languages develop each in their own way, and as far as the expression of genericity is concerned, English and French exploit different possibilities. What is crucial from a synchronic point of view is that both languages have a coherent system, and are equivalent in expressive force. In the next section, I will work out the semantics of definite NPs in such a way that a straightforward explanation of their generic readings becomes available. If definites and indefinites both qualify for the expression of characteristic predication, the differences between English and French become a matter of distribution of labor. Quasi-universal, generic readings in French are constrained by the origin of *des* as a partitive determiner. On the other hand, English does not fully exploit the semantic possibilities of definite NPs in generic contexts.

3 Definite NPs

In the unselective binding framework the question of definite generics has not been studied in much detail. This is not surprising, for in English, and other Germanic languages, these expressions occur only rarely in sentences expressing characteristic predication. But in French and other Romance languages, inductive generalizations are typically expressed by either the indefinite singular, or the definite plural. According to Krifka et al., this implies that the definite article may be used with semantically indefinite NPs in Romance languages. It would indicate the position of the NP in the partition of the sentence, namely that it occurs in the restrictor. In this view, definite NPs in characterizing sentences can also display variable behavior, and are therefore bound by an unselective quantifier. Although it is rather unclear what a semantically indefinite definite NP would be in this context, I take the intuition seriously that in certain contexts definite NPs also get bound variable readings. That requires a more precise description of the relation between bound variable and non-bound variable readings of definites. In order to make this work in the unselective binding framework, we would have to treat definite NPs as variables, but then we also have to make sure that, if the definite does not occur in the restrictor, it gets a definite, rather than an indefinite interpretation:

(40)   a. When Mary receives a letter, she often throws away the envelope.

    b. Quand Marie reçoit une lettre, elle jette souvent l'enveloppe.

In both the English and the French example, the definite NP is interpreted in the matrix, where it gets a non-generic interpretation.[7] This means that definite NPs in general are not semantically indefinite, even in Romance languages. To account for these data in a framework which treats definites as variables, we would have to come up with a version of 'definite' closure, in order to avoid these variables to be captured by existential closure. Although this might be feasible, it is not a very attractive solution, I think. Instead, I will argue that a treatment of definite NPs as context-dependent quantifiers provides a unified analysis of both the generic and non-generic readings of definite NPs.

3.1   Context-dependent quantifiers

Partee's (1987) analysis of definite NPs makes use of the iota-operator as in (41a):

(41)   a. $\| \text{ the king } \|$: $\iota x \, [\text{King(x)}]$
    b. $\| \text{ the king } \|$: $\lambda Q \, [\exists x \, [\forall y \, [\text{King(y)} \leftrightarrow y = x] \wedge Q(x)]]$

The iota-operator combines with an open sentence to give an entity-denoting expression, denoting the unique satisfier of that open sentence if there is just one, and failing to denote otherwise. Montague's generalized quantifier interpretation of definite singular NPs is given in (41b). In both cases there is a uniqueness claim about the referent of the definite NP. So far, this uniqueness claim is formulated with respect to the model as a whole. This seems to be too strong: in many cases we can felicitously use a definite NP like *the king*, even though there are many kings in the universe. Uniqueness should thus be relaxed to the context of use: in the context, there is a unique relevant individual which satisfies the description. Adding such a notion of context dependence leads to the truth conditions in (42a) (cf. Dowty and Brodie 1984; Westerståhl 1984), where the predicate Rel picks out the contextually relevant kings:

(42)   a. $\| \text{ the king } \|$ with respect to an event $e =$
    $\lambda Q \, [\exists x \, [\forall y \, [[\text{King(y)} \wedge \text{Rel}(y, e)] \leftrightarrow y = x] \wedge Q(x)]]$

---

[7]Of course the sentence carries a presupposition of existence, so we quantify over situations in which Mary receives letters that come with an envelope (see Karttunen and Peters 1979 for discussion of the inheritance of presuppositions in complex sentences).

b. $\text{Rel} = \{\langle x, e\rangle |\, x \text{ is relevant in } e\}$

In the event-based semantics used here, predicates always come with an extra Davidsonian argument. This means that the predicate Rel will be interpreted as the set of relevant individuals in a particular event e as in (42b). So definite NPs pick out the individual that is relevant with respect to a certain event e. We can easily account for the behavior of definite NPs at the discourse level by defining them as externally dynamic context-dependent quantifiers, as in (43):

(43) $\|\mathbf{the}\ d_1\ P(d_1)\|$ with respect to an event $e =$
$\lambda Q\,[\mathcal{E}d_1\,[\mathcal{A}d_2\,[[P(d_2)\,;\,{\uparrow}\text{Rel}(d_2, e)] \Leftrightarrow {\uparrow}\,[d_2 = d_1]]\,;\,Q(d_1)]]]$
$= \lambda Q\,\lambda p\,[\exists x\,[\forall y\,[[\downarrow\,{}^\vee P(y) \wedge \text{Rel}(y, e)] \leftrightarrow y = x] \wedge\,\downarrow\,{}^\vee Q(x)\,\wedge\,\{x/d_1\}^\vee p]]$

Using this interpretation schema, (44) translates as (44a), which reduces to (44c) via (44b):

(44) The dean came in. She sat down and talked for an hour.

a. $[\mathcal{E}d_1\,[\mathcal{A}d_2\,[[{\uparrow}\,\text{Dean}(d_2)\,;\,{\uparrow}\,\text{Rel}(d_2, e)] \Leftrightarrow {\uparrow}\,[d_2 = d_1]]\,;$
${\uparrow}\,\text{Come-in}(d_1)]\,;\,\text{Sit-down}(d_1, e)\,;\,\text{Talk}(d_1, e)]$

b. $[\lambda p\,[\exists x\,[\forall y\,[[\text{Dean}(y) \wedge \text{Rel}(y, e)] \leftrightarrow y = x] \wedge$
$\text{Come-in}(x)\,\wedge\,\{x/d_1\}^\vee p]]\,;\,\text{Sit-down}(d_1, e)\,;\,\text{Talk}(d_1, e)]$

c. $[\lambda p\,[\exists x\,[\forall y\,[[\text{Dean}(y) \wedge \text{Rel}(y, e)] \leftrightarrow y = x] \wedge$
$\text{Come-in}(x) \wedge \text{Sit-down}(x) \wedge \text{Talk}(x)\,\wedge\,\{x/d_1\}^\vee p]]]$

The event variable remains free in these representations; I assume that it gets a value from the context of use. The treatment of anaphoric relations extends to quantificational constructions as usual. In the representations given below, I will write the $N^e(d_1)$ as shorthand for $[\mathcal{E}d_1\,[\mathcal{A}d_2\,[[N(d_2)\,;\,{\uparrow}\,\text{Rel}(y, e)] \Leftrightarrow {\uparrow}\,[d_2 = d_1]]]]$ with respect to a certain event e. The treatment of definite NPs as context-dependent quantifiers gives us a straightforward interpretation of the examples in (40), repeated here as (45):

(45)   a. Quand Marie reçoit une lettre, elle jette souvent l'enveloppe.
       b. When Mary receives a letter, she often throws away the envelope.
       c. ${\uparrow}\,\text{OFTEN}\,[\lambda e\,\exists x\,\text{Letter}(x) \wedge \text{Receive}(\text{mary}, x, e)]$
          $[\lambda e\,\exists x\,\text{Letter}(x) \wedge \text{Receive}(\text{mary}, x, e)\,\wedge$
          $\mathbf{the}\ y\ \text{Envelope}^e(y, x) \wedge \text{Throw-away}(\text{mary}, y, e)]$

Adverbs of quantification denote relations between sets of events as usual and definite NPs are not treated as variables, but as context-dependent quantifiers. The desired interpretation of (43) now falls out naturally, and we do not need a special rule of 'definite closure'. There is no reason to assume a difference in behavior between definite NPs in

English or in Romance languages. Also, in neither of these examples would we intuitively characterize the definite NP as generic.

3.2 Definite generics

The interesting step is of course to see how this interpretation allows us to account for generic readings of definite NPs. I will treat such cases as an extension of examples like (46a), which involve a singular NP:

(46) a. When the king dies, he is usually succeeded by his son.
      b. $\uparrow$ USUALLY [λe **the** x King$^e(x) \wedge$ Die(x, e)]
         [λe **the** y Son$^e(y, x) \wedge$ Succeed(y, x, e)]

The predicate *to die* in (46) is once-only, so the set of events in which the individual which is the unique king in a particular context dies is a singleton set of events. As a result, quantification over events in which the unique king in that context dies is equivalent to quantification over pairs $\langle x, e \rangle$ of an individual x and an event e such that x is the unique king in e and x dies in e. This boils down to quantification over kings that die, and the generalization stated is that in most of these cases the king is succeeded by his son.

Similar observations can be made about examples which concern individual-level predicates such as (47):

(47) a. When the king has blue eyes, he is often popular.
      b. $\uparrow$ OFTEN [λe **the** x King$^e(x) \wedge$ Have-blue-eyes(x, e)]
         [λe Popular(x, e)]

Individual-level predicates are always once-only, so quantification over events is equivalent to quantification over pairs $\langle x, e \rangle$ of the unique king x in the context who is such that he has blue eyes in e. Given that individual-level predicates are only loosely tied to particular occasions, and we assume that every king is unique with respect to a particular context, this boils down to quantification over blue-eyed kings.

The context-dependent character of definite NPs remains strongly present, even if we embed the NP under quantification. Although it is reasonable to assume that for every individual involved we can find a context with respect to which the uniqueness condition is satisfied, the fact that there is such an additional condition accounts for the difference felt between (46) and (47) on the one hand and (48a, b) on the other hand, where no extra assumptions are necessary for the set of individuals that satisfy the common noun to become available:

(48) a. When a king dies, he is usually succeeded by his eldest son
      b. When a king has blue eyes, he is often popular

The contrast extends to plural generic NPs as in the following pair of sentences, discussed by Hawkins (1978):

(49) a. Generals usually get their way
    b. The generals usually get their way

In (49a), the most probable intention of the speaker is to refer to all generals. The indefinite can cover all those in existence and all those that will exist. The definite article in (49b) singles out the object mentioned against the background of a more inclusive whole, which could be something like 'officers' or 'people in government'.

As far as French is concerned, we observe that the difference between the singular indefinite and the singular definite NP is preserved. However, the contrast between (49a) and (b) disappears, and we only use the context-dependent definite NP to express generalizations:

(50) Le plus souvent, les généraux imposent leur volonté
    Usually, the generals impose their will

The context-dependent character of definite NPs is less strongly present here. It seems easy to rely on a flexible notion of contextual relevance which allows every individual that satisfies the predicate to satisfy the uniqueness condition with respect to one context or another. Such a weak interpretation of the uniqueness condition blurs the distinction between pseudo-binding with definite and indefinite NPs, and allows French to get by without a generic indefinite plural. Arguably, such a weak notion of context dependency does not apply to English, because there we have the possibility of switching to an indefinite NP if we want to capture the full general reference of the NP.

3.3 Definite disclosure

For those who wish to express bound variable readings as variables actually bound in the semantics, I will show that we can reflect this meaning effect in the semantics if we adopt suitable type-shifting operations. As Partee (1987) points out, applying the BE operator to a definite NP gives us the singleton set of the unique individual satisfying the description if there is one, the empty set otherwise:

(51)   a. $\text{BE}(\text{THE}_{sg}(\text{P})) = \text{P}$ if $|P| \leq 1$[8]
      b. $\lambda x \, [\forall y \, [\text{King}(y) \leftrightarrow y = x]]$

If we relax the uniqueness presuppositions to uniqueness-in-context, as we have argued for above, the type-shifting operator BE gives us for every context a singleton set of individuals satisfying the description of the common noun within that context. More generally, this operator

---

[8] Where THE is the interpretation rule for *the*, i.e.:

$\lambda P \, \lambda Q \, [\exists x \, [\forall y \, [P(y) \leftrightarrow y = x]] \wedge Q(x)]$

Remember that Partee uses the Russellian interpretation of the definite description, so there is no relativization to the context as yet. We will see in a moment how her analysis works out if we assume contextually dependent uniqueness.

gives us the set of pairs $\langle x, e \rangle$ such that x is the unique x satisfying the definite description in e. Conjunction with an identity statement thus does not only give us existential disclosure, it also provides for 'definite disclosure':

(52) When the king has blue eyes, he is often popular

   a. $[\lambda e \downarrow [\mathcal{E}d_1 [\mathcal{A}d_2 [[\uparrow \text{King}(d_2) \, ; \, \uparrow \text{Rel}(d_2, e)] \leftrightarrow$
      $\uparrow [d_2 = d_1]] \, ; \, \uparrow \text{Have-blue-eyes}(d_1, e)]]]$

   b. $[\lambda y \, \lambda e \downarrow [[\mathcal{E}d_1 [\mathcal{A}d_2 [[\uparrow \text{King}(d_2) \, ; \, \uparrow \text{Rel}(d_2, e)] \leftrightarrow$
      $\uparrow [d_2 = d_1]] \, ; \, \uparrow \text{Have-blue-eyes}(d_1, e)]] \, ; \, \uparrow [d_1 = y]]]$

   c. $[\lambda y \, \lambda e \, [\exists z \, [\forall x \, [[\text{King}(x) \wedge \text{Rel}(x, e)] \leftrightarrow x = z] \wedge$
      $\text{Have-blue-eyes}(y, e) \wedge y = z]]]$

   d. $[\lambda y \, \lambda e \, [\forall x \, [[\text{King}(x) \wedge \text{Rel}(x, e)] \leftrightarrow x = y] \wedge$
      $\text{Have-blue-eyes}(y, e)]]$

   e. $[\lambda y \, \exists e \, [\forall x \, [[\text{King}(x) \wedge \text{Rel}(x, e)] \leftrightarrow x = y] \wedge$
      $\text{Have-blue-eyes}(y, e)]]$

   f. $[\lambda y \, [\text{King}(y) \wedge \text{Have-blue-eyes}(y)]]$

(52a) is the DMG representation of the restrictor on the quantifier. In (52b) this formula is dynamically conjoined with an identity statement. This reduces in the regular way to (52c), and yields the set of pairs $\langle y, e \rangle$ such that y is the unique king in the context and y has blue eyes in e (52d). Because the individual-level predicate carries a uniqueness presupposition, this is equivalent to quantification over the set of blue-eyed kings that are unique with respect to a certain context (52e). Assuming that such a context can be found for every king, we end up expressing quantification over blue-eyed kings (52f).

Just like in English existential disclosure is used for both indefinite singulars and bare plurals, French allows definite disclosure with both singulars and plurals, without any perceptible semantic role for the plurality marker (cf. Kleiber 1990: 25). We can thus account for (53) in the following way:

(53) En général, les chats intelligents ont les yeux verts.
    In general, DEF-PL cats intelligent have eyes green

   a. $[\lambda e \, \exists X \, [\textbf{the } d \, [\uparrow \text{Cat}^e(d) \, ; \, \uparrow \text{Intelligent}(d, e) \, ; \, \uparrow X(d, e)]]]$

   b. $[\lambda y \, \lambda e \, \exists X \, [\textbf{the } d \, [\uparrow \text{Cat}^e(d) \, ; \, \uparrow \text{Intelligent}(d, e) \, ; \, \uparrow X(d, e)]]$
      $; \uparrow [d = y]]$

   c. $[\lambda y \, \lambda e \, \exists X \, [\forall x \, [[\text{Cat}(x) \wedge \text{Intelligent}(y, e) \wedge \text{Rel}(x, e)] \leftrightarrow$
      $x = y] \wedge X(y, e)]]$

   d. $[\lambda y \, \exists e \, \exists X \, [\forall x \, [[\text{Cat}(y) \wedge \text{Intelligent}(y, e) \wedge \text{Rel}(x, e)] \leftrightarrow$
      $x = y] \wedge X(y, e)]]$

   e. $[\lambda y \, [\text{Cat}(y) \wedge \text{Intelligent}(y) \wedge X(y, e)]]$

The restriction on the quantifier, given in (53a), is recovered by association with focus, just as in (33) above. The main difference is that we have a set of events the characterization of which involves a definite NP, whereas the one in (33) involved an indefinite NP. But the relation between individuals and events is the same, because of the once-only character of the predicate: X stands for a suitable alternative to having green eyes (e.g. having blue eyes), so it will certainly be an individual-level predicate. In (53b), an identity statement is dynamically conjoined with the original formula. This leads to disclosure of the definite NP, in such a way that we end up with the set of events in which the unique contextually defined cat is intelligent. Given the once-only character of the predicate, this is definitely equivalent to the set of pairs $\langle x, e \rangle$ given in (53c), which can be reduced to (53d) because the predicate is once-only and there is a one-one correspondence between the set of events and the set of individuals which are the unique relevant individual in that event. We can reduce (53d) further to (53e) if we choose to abstract away from the event variable altogether.

The comparison of (52e)/(53e) with (32d)/(33d) makes it understandable why Krifka et al. (1995) claim that generic definite NPs in Romance can be considered as semantically indefinite NPs. Of course we can only infer (52d) and (53d) from (52c) and (53c) if we appeal to the once-only character of the predicate involved, and further reduction to (52e) and (53e) is also dependent on assumptions about the real world, so even in this semantic approach we crucially need to appeal to pragmatic considerations in order to obtain the right set of individuals to quantify over.

Again, the main difference between definite disclosure and pseudo-binding is where to locate the burden of the explanation: in the semantics or in the pragmatics. Technically, both options are available. Given that we need the pragmatic constraints on uniqueness of event variables anyhow, I prefer to treat bound variable readings of both indefinite and definite NPs as pseudo-binding and preserve a unified analysis of adverbs of quantification as generalized quantifiers over events.

4 Conclusion

I conclude that the interpretation of indefinites as dynamic existential quantifiers allows us to treat *des* N as a regular indefinite as far as discourse and donkey anaphora are concerned. English and other Germanic languages have a preference for the use of indefinite NPs in characteristic generic sentences. Pseudo-binding makes it possible to fix certain indefinite NPs as topics, which determine what the quantification is about. The NPs which undergo pseudo-binding are the only ones which can be properly called generic NPs. This approach accounts for the generic character of the indefinite NPs in (54a) and (54b):

(54) a. An Italian usually drinks wine at dinner.
  b. Italians usually drink wine at dinner.
  c. The Italian usually drinks wine at dinner.
  d. The Italians usually drink wine at dinner.

Although a similar operation of pseudo-binding could be used to turn (54c) and (54d) into generic sentences, this is not the usual way to express characteristic predication in English. English thus does not fully exploit the semantic possibilities of definite NPs in generic contexts.

Not all indefinite NPs in all languages undergo pseudo-binding. As we have seen, French and other Romance languages do allow this for singular indefinite NPs (55a), but indefinite plurals do not usually occur as the topic of a generic sentence (55b):

(55) a. Un Italien boit généralement du vin à table.
  b. *Des Italiens boivent généralement du vin à table.
  c. L'Italien boit généralement du vin à table.
  d. Les Italiens boivent généralement du vin à table.

Instead, Romance languages seem to have a preference for the use of definite NPs for the expression of inductive generalizations. (55c) has about the same status in French and English, but (55d) is clearly the typical way to express characteristic predication. The difference between Germanic and Romance languages then reduces to a preference for pseudo-binding with indefinite or definite NPs.

We can speculate that there is an overall preference for the expression of genericity by means of indefinites rather than definites because the relation between the NP-denotation and the CN-denotation is more straightforward in the case of indefinites. Pseudo-binding of definite NPs is dependent on uniqueness-in-context and is as such a more indirect way of associating sets of events with the correlating sets of individuals. We would then predict that only if pseudo-binding of indefinite NPs is somehow prohibited for independent reasons, languages would use definite NPs to express genericity.

References

Auger, J. 1993. Syntax, Semantics, and *ça*: on Genericity in Colloquiual French. *The Penn Review of Linguistics* 17, 1–12.

Carlier, A. 1989. Généricité du Syntagme Nominal Sujet et Modalités, *Travaux de Linguistique* 19, 33–56.

Chierchia, G. 1992. Anaphora and Dynamic Binding. *Linguistics and Philosophy* 15, 111–183.

Dekker, P. 1993. *Transsentential Meditations: Ups and Downs in Dynamic Semantics*. Doctoral dissertation, University of Amsterdam.

Dowty, D. and B. Brodie 1984. The Semantics of "Floating" Quantifiers in a Transformationless Grammar. In *Proceedings of WCCFL* 3, 75–90. Stanford: Stanford Linguistics Association.

Galmiche, M. 1986. Référence Indéfinie, Evénements, Propriétés et Pertinence. In J. David et G. Kleiber, eds. *Déterminants: Syntaxe et Sémantique*, 41–71. Paris: Klincksieck.

Groenendijk, J. and M. Stokhof 1991. Dynamic Predicate Logic. *Linguistics and Philosophy* 14, 39–100.

Groenendijk, J. and M. Stokhof 1992. Dynamic Montague Grammar. In: L. Kálmán and L. Polós, eds. *Papers from the Second Symposium on Logic and Language*. Budapest.

Hawkins, J. 1978. *Definiteness and Indefiniteness: a Study in Reference and Grammaticality Prediction*. London: Croom Helm.

de Hoop, H. and H. de Swart 1990. Indefinite Objects. In R. Bok-Bennema and P. Coopmans, eds. *Linguistics in the Netherlands* 1990, 91–100. Dordrecht: Foris.

Karttunen, L. and S. Peters 1979. Conventional Implicature. *Syntax and Semantics* 11, 1–56. New York: Academic Press.

Kleiber, G. 1990. *L'article* LE *Générique. La Généricité sur le Mode massif*. Genève: Droz.

Kratzer, A. 1995. Stage-level and Individual-level Predicates. In G. Carlson and F. Pelletier eds. *The Generic Book*, 125–175. Chicago: University of Chicago Press.

Krifka, M. 1995. Focus and the Interpretation of Generic Sentences, In G. Carlson and F. Pelletier eds. *The Generic Book*, 238–264. Chicago: University of Chicago Press.

Krifka, M., F. Pelletier, G. Carlson, A. ter Meulen, G. Chierchia and G. Link 1995. Genericity: an Introduction. In G. Carlson and F. Pelletier eds. *The Generic Book*, 1–124. Chicago: University of Chicago Press.

Lewis, D. 1975. Adverbs of Quantification. In E. Keenan, ed. *Formal Semantics*, 3–15. Cambridge: Cambridge University Press.

Milsark, G. 1974. *Existential sentences in English*. Doctoral dissertation, MIT.

Partee, B. 1987. Noun Phrase Interpretation and Type-Shifting Principles. In J. Groenendijk, D. de Jongh, and M. Stokhof, eds. *Studies in Discourse Representation Theory and the Theory of Generalized Quantifiers*, 115–143. Dordrecht: Foris.

Rooth, M. 1985. *Association with focus*. Doctoral dissertation, University of Massachusetts at Amherst.

de Swart, H. 1991. *Adverbs of Quantification: a Generalized Quantifier Approach*. Doctoral dissertation, University of Groningen. Published 1993, New York: Garland.

de Swart, H. 1995. Position and Meaning: Time Adverbials in Context. Ms., University of Groningen.

de Swart, H. to appear. Quantification over Time. In J. van der Does and J. van Eijck eds. *Quantifiers, Logic, and Language*, Stanford, CSLI Publications.

Vallduví, E. 1992. *The Informational Component*. New York: Garland.

Westerståhl, D. 1984. Determiners and Context Sets. In J. van Benthem and A. ter Meulen, eds. *Generalized Quantifiers in Natural Language*, 45–72. Dordrecht: Foris.

Index